The Classical Gardens of Shanghai

RAS CHINA in SHANGHAI

RAS CHINA in SHANGHAI

In 1857 a small group of British and Americans seeking intellectual engagement in a city dedicated to commerce established the Shanghai Literary and Scientific Society. Within a year the organization was granted affiliation with the Royal Asiatic Society of Great Britain and Ireland in London and the North China Branch of the Royal Asiatic Society was born. The society was reconvened in Shanghai in 2007.

The RAS China in Shanghai Series of China Monographs, published in association with Hong Kong University Press, is designed to reflect the vibrancy as well as the wide research interests and contacts of the society and to provide a forum for its members and associates to publish their research interests.

Series editor: Paul French

Other titles in the RAS China in Shanghai series:

The Happy Hsiungs: Performing China and the Struggle for Modernity
Diana Yeh

Knowledge Is Pleasure: Florence Ayscough in Shanghai
Lindsay Shen

Lao She in London
Anne Witchard

Mu Shiying: China's Lost Modernist
New Translations and an Appreciation by Andrew David Field

The Classical Gardens of Shanghai

Shelly Bryant

HKU PRESS
香港大學出版社

Hong Kong University Press
The University of Hong Kong
Pok Fu Lam Road
Hong Kong
https://hkupress.hku.hk

ISBN 978-988-8208-81-4 (*Paperback*)

British Library Cataloguing-in-Publication Data
A catalogue record for this book is available from the British Library.

Digitally printed

Contents

Acknowledgements

Special thanks are due to Tongji University's Yu Lan for generously offering her time, patience, resources, and insights into gardening culture during the research phase of this monograph, and to Shanghai International Studies University's Sun Li for introducing me to classical Chinese gardens and laying a foundation of knowledge on which greater love and appreciation for gardening culture could grow over the years. The assistance Sun Li and Loh Nyuk Fong offered in proofreading the drafts of the translations in this manuscript was invaluable.

The Royal Asiatic Society in Shanghai and Beijing has made this publication possible through their financial support. In addition, many members of the society have been of great help along the way, especially Miho Kinnas, Kate Baker, Susie Gordon, Lindsay Shen, and Linda Johnson, who have all travelled with me to visit the gardens discussed in this monograph on various occasions. Miho Kinnas, Lindsay Shen, Susie Gordon, and Spencer Dodington offered further assistance in serving as first readers for the manuscript, and Susie has generously made her photos available for the monograph.

I am particularly indebted to the Editorial Advisory Committee, to Paul French, who has been a tireless editor, offering insights and comments along the way and helping to tie up loose ends, and to Yuet Sang Leung at Hong Kong University Press. The help of such a top-notch editorial team is greatly appreciated.

Unless otherwise noted, the Chinese poetry cited in this monograph is quoted from the Shanghai Municipal Government's collection, which can be accessed online at www.shtong.gov.cn/node2/node2245/node69854/node69866/node69882/index.html.

All English versions found in this monograph, unless otherwise noted, are the author's own translations.

1
Introduction: Shanghai's Gardens in Context

> The wise take pleasure in water, and the kind find happiness in a mountain.
>
> —Confucius[1]

Conventional wisdom in the Chinese classical gardening tradition says that it takes three hundred years to build a garden. The appreciation of a garden is likewise not something that happens in a short space of time, but instead is cultivated over the course of a lifetime. A scholar's garden involves the cultivation of the land, the self, the family, and the surrounding society and culture, all in the confines of one household's walled residential compound. The role that the garden played in the daily lives of families in times past was to create spaces for reading, entertainment, and enjoyment. The residential portions were meant to be secluded, reading areas quiet, entertainment spaces convenient, and enjoyment sections relaxing.[2]

1. Lou Qingxi, *Chinese Gardens,* trans. Zhang Lei and Yu Hong (Beijing: China Intercontinental Press, 2003), 126. Hereafter cited as Lou 2003.
2. Ibid., 56.

Viewing a garden is not meant to be a passive activity. The visitor is expected to enter the garden not only physically, but also intellectually and emotionally, and to respond to what she or he encounters on all three levels. The viewer should seek to put him or herself into the art when visiting a garden.[3] A garden must be experienced with the senses, with the mind, and with the heart, the last of which is the true goal of garden construction. Ji Cheng (計成) writes, 'There is no definite way of making the most of scenery; you know it is right when it stirs your emotions.'[4] Engaging the senses—which Lou Qingxi (樓慶西) terms 'the image, or realm of the substance' in contrast with 'the meaning, or artistic/emotional realm'[5]—is the most obvious and accessible level for experiencing a garden, while it could be argued that responding emotionally is not possible without first engaging in an intellectual response. Lou states that the real appreciation of a garden's beauty comes from an understanding of the general artistic realm behind the scenery presented there, which leads to an understanding of the philosophy and view of life it represents.[6] This intellectual response depends entirely on the viewer's familiarity with the Chinese tradition of gardening.

This monograph seeks to provide information that will help readers interact with Shanghai's gardens intellectually, in hopes that this intellectual engagement will open up the possibility for emotional engagement as well.

Chinese landscaping is a fundamentally different endeavour from Western gardens, where wide vistas are laid out, usually

3. 彭一剛：《中國古典園林分析》（北京：中國建築工業出版社，1986），8. Hereafter cited as Peng.
4. Ji Cheng, *The Craft of Gardens*, trans. Alison Hardie (Shanghai: Shanghai Press and Publishing Development Company, 2012), 120.
5. Lou 2003: 125.
6. Ibid., 126.

symmetrically over a large piece of land, and floral scenes are foregrounded. In a classical Chinese garden, the symmetry is not spatial but philosophical, and the grounds are partitioned so as to create winding paths that reveal the whole in piecemeal fashion, employing screens to create and frame miniature landscapes, resulting in a layered effect that seems to expand the relatively small confines within the garden's walls.

In English writings, the term 'classical Chinese garden' generally refers to the gardens found in the Jiangnan region (江南, the southern part of the Yangtze River Delta), with Suzhou serving as the long-established centre of the landscaping techniques found in this style of garden. The terms private, scholar's, poet's, southern, classical, Suzhou, Chinese, and Jiangnan are all used interchangeably to denote the gardens that grew up in the region, forming one of three distinct styles of gardens in Chinese tradition, alongside imperial and landscape gardens. Temple gardens, which are sometimes treated as a fourth classification, generally follow the Jiangnan style in smaller grounds and imperial gardens in larger compounds.[7]

A typical Suzhou garden is composed of four elements—waterways, rockeries, buildings, and plants—with walkways and corridors arranged to connect them while also situating guests to the grounds at the most advantageous viewing positions. These four types of features are arranged to enhance the natural topography and are arranged in relation to one another in such a way as to create a variety of views from many different angles. Each feature can serve simultaneously as a part of a view, a point from which to observe other scenes in the grounds, and a partition to separate sections of the garden.

7. Ibid., 6.

The southern-style landscaping tradition reaches back as early as the fourth century BC in the Chu Kingdom,[8] but the techniques by which the grounds are designed and built in a classical garden were not written down in systematic fashion until the Ming dynasty designer, Ji Cheng, composed his 1631–34 volume *The Craft of Gardens*.[9] Since this is the period during which Shanghai's gardening culture flourished, Ji is a good place to begin a study of Shanghai's gardens. Even in contemporary writing about Chinese gardens, it is not uncommon for a significant work in Chinese publications to be overlooked for decades by both Chinese and foreign scholars,[10] and this was even more true of older texts. Ji Cheng's handbook on landscaping techniques was 'lost' for generations until Japanese scholars, who had long relied on the work, reintroduced it to China in 1931.[11] Ji's text was unusual for its time, as most traditional writings on gardens were records 'focused on concrete particulars [of a specific garden], and there is an enduring reluctance to operate at the level of treatises on general principles.'[12] Probably from a humble background, Ji Cheng describes himself as well-travelled, though nearly all of his design work was done in his native province of Jiangsu. Ji Cheng's reputation was apparently established not by rights of birth, but by his skill. The goal in his philosophy of design was to make a visitor feel that she or he had explored the whole of Jiangnan, even if the actual journey had only covered a path of a mile or so.[13]

---

8. Ji, 15.
9. Ibid., 11.
10. Stanislaus Fung, 'Longing and Belonging in Chinese Garden History', *Perspectives on Garden Histories* 1999: 205–19. Print, 205.
11. Ji, 12.
12. Fung, 215.
13. Ji, 39.

Ji's summary of the techniques of garden building focus on the four elements mentioned above. In the preface to his guidebook for garden design, he writes:

> To make a garden here, one should not only pile up rocks to emphasize the height, but excavate the earth to increase the depth in proportion with the tall trees scattered on the hillside here, with their roots curled around sheer rocks just as in a painting. Following the course of the stream we should construct pavilions and terraces whose reflections will be scattered on the surface of the pond, with winding gullies and flying galleries leading on from them, so that people will be taken beyond anything they could have imagined.[14]

From this brief summary, it is evident that rocks and waterways play the more prominent role in a Suzhou garden, with buildings and plants serving to enhance the layout of the other two elements. This is fitting with the conventional concept of a classical garden functioning as a series of landscape paintings, termed *shanshui* (山水, literally 'mountains and water/rivers') paintings in Chinese. The rockeries serve as miniaturized mountains with waterways surrounding them, setting up a conceptual symmetry between permanence (mountain) and change (water) that represents balance and harmony in Chinese aesthetics, and the relationship between the two is always foregrounded. 'The waters follow the hills, and the hills are brought to life by the waters' in a Chinese garden.[15] Plants serve as forestation in the miniaturized landscape, and the buildings function simultaneously as parts of the landscape, framing devices through which to view the grounds, and

14. Ji, 37.
15. Chen Congzhou, *On Chinese Gardens* (Shanghai: Shanghai Press and Publishing Development Company, 2008), 16.

places of rest to break up the walk. The views in a garden serve as a 'concentration of scenery', or what Lou Qingxi calls a 'world in a teapot',[16] that is meant to make the viewer feel that the grandeur of nature has been brought near.

The literal translation of the common term for garden building in Chinese is 'digging ponds and piling mountains' (挖湖堆山, *wa hu dui shan*).[17] Ji Cheng gives extensive attention to both aspects of garden building. When laying rocks, it is most important to begin with an overall picture in mind, leaving the details for a later stage.[18] The process begins with using wooden posts as models to estimate the height and length of the hills, and also to test the firmness of the ground on which the miniature mountain range will be built. A block and tackle is then set up, using ropes to lift the stones. The base of the mountain should consist of large, rough stones that can be laid so as to completely hide the wooden pillars around which the hill will be built. These pillars, serving as the 'bones' of the mountain, should be driven far into the ground. When the large, solid rocks have been laid, smaller, jagged pieces can be added as the mountain gains height, bringing texture and variety to the hill's layout. Ji Cheng adds that 'thin' and 'riddled' rocks are naturally impressive, while smoother pieces have to rely on the garden master's ability to position them perfectly in order to create an imposing view.[19]

The two types of rocks most commonly used in gardens are huang and Taihu stones. The base is easier to lay for huang formations, while it is difficult to situate the top of the structure. The reverse is true of Taihu rockeries. Huang stones are solid

16. Lou 2003, 20.
17. Ji, 22.
18. Chen, 66.
19. Ji, 104.

Figure 1.1

A Taihu stone with plants, forming a typical scene in a classical Chinese garden (© 2015 Song Jie)

and vigorous, with a hint of hollowness and flexibility detectable. In contrast, Taihu stones emphasize hollowness. Each type also comes with disadvantages. For instance, huang rocks lack a sense of change, while Taihu's stones can appear fragmentary and scattered.[20] An approach that carefully balances the strengths and weaknesses of these two forms of rocks is fundamental to good garden design.

When choosing rocks, one should seek pieces without cracks, strong enough to be piled up, and with jagged edges that can be fitted together. Even ordinary rocks can be beautiful when stacked

20. Chen, 66.

together properly. 'Rocks,' Ji Cheng writes, 'are not like plants or trees; once gathered, they gain a new lease on life'.[21]

If one finds an especially beautiful rock, it may be displayed on its own in a garden, and individual stones are often positioned as the centrepiece of a view within the grounds. For the Confucian thinker, the contemplation of rocks was traditionally an aid to cultivating character, as rocks are sturdier than trees or plants, and therefore represent solid character.[22] Ming rockeries, characterized by strength and massiveness, aimed to evoke contemplation. During the Qing dynasty, there was a push for more refinement and sophistication in rock formations, but it often resulted in structures that were too fragile to last.[23] Taihu stones have long been favoured for this sort of individual display. These stones were traditionally drawn from Taihu, usually from the edges of the lake near Dongting Hill or from Xiaoxia Bay.[24] The ideal stone is firm and glossy, full of hollow spaces, holes, twists, and intriguing grooves. They may be of a variety of shades of white, grey, or black. A good stone will give out a faint sound when tapped.[25]

Ming rockeries, of which there are several examples in Shanghai's gardens, were generally kept quite simple. They included steps, main peaks, flat terraces, ravines, and caverns in their arrangements, laid out in a myriad of styles. The two main classifications were 'open', with ravines hewn out of hills, and 'closed', in which the main peaks are piled one above another.[26]

Surrounding the various rockeries in a classical garden will be a system of waterways, the juxtaposition of these two elements

21. Ji, 112.
22. Ibid., 134, note 165.
23. Chen, 66.
24. Ji, 112.
25. Ibid., 112.
26. Chen, 67.

representing a yearning for virtue and wisdom, even as it mimics natural landscapes.[27] As a general rule, roughly 30 percent of the grounds are dedicated to water. If the mountains are the skeleton of the garden, the waterways are the arteries, and in them one feels the pulse of the place. The ponds and canals are the yin to the mountains' yang.[28] In Zheng Yuanxun's (鄭元勳) 1635 foreword to Ji Cheng's volume, he refers to streams that appear as an 'undulating ribbon' winding through the artificial hills.[29] This is the ideal that the waterways in a garden aim for.

In the earliest gardens, mountains were the most important elements, favoured for the representation of the houses of the gods. As time wore on, water was more heavily emphasized, leaving the mountains to the immortals. The purpose of the garden was to escape the confines of reality, and the canals in a garden were seen as a pathway for that journey, much like the fisherman escaped to the utopian society beyond the peach orchard in Tao Yuanming's classic 'The Land at the End of the Peach Grove'.[30] The emphasis on reconnecting with nature meant that pools and canals should appear as they might in the wild, winding in paths shaped by the topography of the land, asymmetrical and seemingly random.[31]

The ponds should be of irregular shape like natural bodies of water.[32] More expansive ponds or lakes can be used to showcase the change of seasons, reflecting the colour changes surrounding it.[33] As much as possible, gardens should be constructed to

27. Cai Yanxin, *Chinese Architecture,* trans. Andrea Lee, Selina Lim, and David Gu (Beijing: China Intercontinental Press, 2010), 126.
28. Ji, 22.
29. Ibid., 31.
30. Peng.
31. Lou 2003; 62.
32. Ji, 51.
33. Ibid., 45.

incorporate natural waterways. Dredging and pumping water in is also acceptable, but the pumps should be hidden from view. In a compound that is surrounded by natural bodies of water, it is advisable to 'borrow' the view, a technique used to great effect in Suzhou's oldest surviving garden, the Surging Waves Pavilion. Though every garden is self-contained, when one can make use of such natural elements surrounding it, the creative employment of windows, doorways, and corridors to frame those views and draw them into the garden is not only acceptable, but commendable. Ji Cheng's sage advice is that vulgar views surrounding the garden should be blocked out and beautiful ones borrowed.[34]

In the Ming dynasty (1368–1644), lakes in imperial grounds were given names such as Beihai, Zhonghai, and Nanhai, *hai* (海) being the term for 'sea'. Building on this tradition, just as the rocks represented miniature mountains in southern gardens, the lakes and ponds stood for oceans.[35] In southern gardens, unlike their northern counterparts, it was necessary that the water in a garden came from a source inside or very near the edges of the walled compound,[36] so the presence of a ready water source was always a primary consideration in choosing a site for gardens in Jiangnan. In a place like Hangzhou, where a major body of water dominates the topography, landscape gardening is seen as a natural activity in which to engage.[37] Throughout the Yangtze Delta area, the ready availability of natural canals means that finding a good waterway to build around was never a very big obstacle to garden designers. The preference for building around waterways in the south also

34. Ibid., 43.

35. Peng, 3.

36. 樓慶西：《中國古建築二十講》（北京：生活・讀・新知三聯書店，2001），167。Hereafter cited as Lou 2001.

37. Peng, 4.

bore a subtle political note in the earliest days, as northerners were plainsmen, good with horses, while southerners were more skilled with boats, and water emphasized their area of strength.[38]

It was preferred that waterways be not too large, and that water not overwhelm the grounds.[39] Bridges were often used to separate the watery expanses,[40] and stone bridges became an iconic representation of Jiangnan over the years. Ideally, bridges should be lower than the pond's banks, so that they will seem to hover on the surface of the water.[41] When water is situated near a building, it should appear to flow out from the roots of the structure. Rocks should be situated so as to create high spaces from which to view the water, and also to make the edges of the pools, ponds, and canals more jagged and their shapes irregular.[42]

Chen Congzhou (陳從周), the renowned professor of Chinese architecture at Tongji University until 2000, makes the observation that the buildings in a garden are where human life is contained, and therefore are the means for generating an artistic sense within the site's natural scenes.[43] Ji Cheng describes the architecture found in a classical garden at great length, including illustrations of the various types of structures, and also the doorways, windows, and balustrades that ornament them. He points to the consideration of views in order to determine the layout of the buildings in a garden as one of the most important things that a designer will do.[44] For residential buildings, Ji advises that one 'should always follow

38. Ibid., 5.
39. Lou 2001, 173.
40. Lou 2003, 62.
41. Chen, 32.
42. Lou 2001, 174.
43. 唐名聲：《海派園林》（上海：文匯出版社，2010），35。Hereafter cited as Tang.
44. Ji, 55.

what is elegant and simple, and take what is most orthodox and straightforward from ancient times.'[45] Painting, decoration, and carvings should always be avoided in residential structures within the grounds.[46]

For other buildings, Ji's theory goes to the opposite extreme. He writes, 'You must search out the unconventional and make sure it is in accord with your own wishes. The trite and conventional should be totally eliminated.'[47] This range of structures, in contrast to the residential buildings, are made for enjoying the grounds, and so the buildings should be laid out so as to both capture and create scenery,[48] and they should 'serve to screen different parts of the grounds, so that they may be revealed at leisure, thus creating a perception of depth within the garden.'[49] The garden's study should not have a strict separation between inside and outside, creating instead a secluded site with easy access to the scenery.[50] Terraces can be built atop a rockery,[51] while pavilions are made to be places for guests to stop and rest. Gazebos, meant for taking advantage of a particular view, are usually situated beside a waterway or plant arrangement, and galleries are spacious, lofty halls for showcasing art or floral arrangements.[52] Penthouses, built against a cliff face, are not complete in themselves, but use the rocks to form part of the structure.[53] The consideration of buildings such as those

---

45. Ibid., 64.
46. Ibid., 64.
47. Ibid., 66.
48. Lou 2003, 57.
49. Ye Shengtao, 'The Suzhou Garden'. Author's translation, original text available online at http://www.5156edu.com/page/07-06-07/24814.html.
50. Ji, 60.
51. Ibid., 69.
52. Ibid., 70.
53. Ibid., 71.

described in detail by Ji Cheng rose to a position of prime importance in garden layout during the Ming dynasty, and Shanghai's historic gardens contain many examples of Ming structures.[54]

Windows and doorways are important parts of the buildings within a garden. 'Wooden walls should have many window-openings so that one can secretly enjoy looking through them into different worlds.'[55] In Suzhou's gardens alone, one may find hundreds of styles of window lattice patterns and window shapes, looking from a distance like flowers drawn on a white piece of paper.[56] The windows, a trademark of the region's gardens, are moulded with black strips and clay, with clean outlines to create sharp images.[57] Balustrades, usually constructed offsite and brought in, should remain simple but elegant. Intricate carvings are not preferred, and religious designs should be reserved for temple gardens.[58]

Carvings in doorjambs should be avoided. Ji Cheng's view of the ideal doorway is that it 'should lead one on to the open spaces and . . . draw one close to the scenery'.[59] Walls should be whitewashed and covered with a mix of paper pulp and lime[60] and they should be free of carvings. Ji Cheng offers two reasons for this

54. Tang, 9. Throughout this discussion, an indication that a particular feature dates from a certain dynastic period (i.e., 'Ming structures') does not necessarily mean that the current brick and mortar building has stood since that time. Rather, following Chinese convention, it refers to the time the building was designed, named, and first built, and/or the style employed in the reconstruction. Unless otherwise stated, it should be assumed that most current structures are rebuildings, as most of the gardens have been through numerous phases of decline and rebuilding throughout their history.
55. Ji, 75.
56. Lou 2003, 66.
57. Ibid., 66.
58. Ji, 86.
59. Ibid., 93.
60. Ibid., 96.

preference, stating that carvings on walls are ugly, but also impractical, as they create spaces for birds to build nests and creepers to grow.[61] Even the plainest sort of wall, a blank white 'mirror wall', can be used to create exquisite scenery by providing a reflective surface for the play of light at various times of day. Open brickwork walls are used freely throughout a garden at any point where there is something worth viewing. These walls create an impression of protecting the scene from outside elements, and also conceal what is waiting so as to allow for a slow revelation of the garden's treasures.[62] A good rule of thumb is that 'walls divide but do not separate'.[63]

Plants are arranged in the grounds so as to emphasize the change of seasons. 'There are few flowers that do not wither, but fresh scenes can be enjoyed all year round',[64] so the use of plants helps to generate a feeling of newness throughout the year and keeps the garden always interesting. Ji Cheng advocates that plants, like rocks and water, should be used in a way to imitate natural scenery. He writes, 'Don't commit crimes against the hills and forests. A man of sensibility will never treat them irreverently.'[65] Lou Qingxi, professor of the history and theory of ancient Chinese architecture, notes that a garden's plants should be arranged with consideration of whether rocks or earth are more important in a given scene. If rocks, then plants should be kept to a minimum, while more plants can help emphasize the earth in a scene.[66] Chen Congzhou, in his extensive studies and garden restoration works, notes that specific types of plants were often emphasized in a given

61. Ibid., 90.
62. Ibid., 96.
63. Lou 2003, 59.
64. Ji, 120.
65. Ibid., 52.
66. Lou 2001, 173.

garden, such as pines in Liu Yuan, plums and pines in Yi Yuan, and bamboos in the Surging Waves Pavilion—three of Suzhou's classical gardens—but that today, different species are generally mixed together in most restored sites, diminishing the individual character of the original garden.[67] Ming dynasty writer and painter Wen Zhenheng (文震亨) writes about the use of plants in a Jiangnan garden: 'Plants and trees should not be mixed together, but one should plant them in the right places, so that there is always something to look at throughout the four seasons, as though one were living within a painting.'[68] In southern gardens, some commonly used plants include peach blossoms and willows for the spring, maples and osmanthus for the autumn, and pines, cypress, banana trees, and bamboo to provide greenery all year. Lotus and water lilies are often used in ponds, with the lotus planted so as to be viewed from a distance and the lilies standing closer to the banks. Bonsais are frequently scattered throughout the grounds as a sort of finishing touch.[69]

Chen reminds his readers of one key difference between not only Chinese and Western gardens, but also scholar's and botanical gardens in China. In a Jiangnan garden, the quantity of plants is not significant. Rather, it is the creation of views that matters, and laying out the small space of the grounds so as to capture a wide variety of scenes that is of primary consideration.[70] The result is that a very small space is enlarged in the mind of the visitor, in part because the viewer is drawn into the scene and invited to walk through it. Lou says, 'With every step, the view will change'

67. Chen, 21.

68. Quoted in Alison Hardie, *Chinese Garden Pleasures* (Shanghai: Shanghai Press and Publishing Development Company, 2014), 109.

69. Lou 2003, 69.

70. Chen, 22.

in a classical garden,[71] and so the space seems to be expanded. At the same time, the miniaturized scenes are meant to be more than merely an imitation of natural landscapes. The designer will purposely employ exaggeration, aiming not merely to represent nature, but to express and inspire certain feelings about it.[72] In ideal situations, 'a mood [is] created from nature but surpassing nature, a mood unique to Chinese culture'.[73]

Though various stages of the gardening tradition had begun much earlier in China, it was during the Song dynasty (960–1279) that the southern style began to gain widespread popularity. During this period, economic development was on the wane, but the arts, including landscape designing, flourished. The Song dynasty was the period that saw the rise of *wenren* (文人, 'cultured person' or literati), a class who was mainly responsible for the development of the scholar's or poet's gardens of the Jiangnan region.[74] Importantly, as Shanghailander and twentieth-century translator of Chinese poetry, Florence Ayscough notes in her essay on Chinese gardens, 'The only aristocracy which has ever existed in China, excepting members of the Imperial clan, was an aristocracy of the brains.'[75] During the period when this cultured, educated elite was rising, a theory of the arts was formulated, a mindset that eschewed realist representation in favour of meaning.[76] Landscaping techniques had an obvious appeal for the literati, since '[i]n contrast with natural scenery, the beauty of Chinese gardens lies in their combining

71. Lou 2001, 171.
72. Peng, 8.
73. Lou 2003, 26.
74. Peng, 3.
75. Florence Ayscough, *A Chinese Mirror* (Boston and New York: Houghton Mifflin Company, 1925), 126.
76. Peng, 3.

culture and art into one',[77] and while one did not need to be rich to construct a garden, he did need to be cultured.[78] It was this framework that allowed for the phenomenal growth in gardening culture, moving from the more linear styles of the imperial garden to an approach that embraced the 'random orderliness' of nature.[79]

It was also during the Song dynasty that contrasts and balances began to take clearer form in landscape design. The term *yuanlin* (園林, 'garden') is made up of two characters, the first which emphasizes the walled compound (i.e., artificial) and the second meaning 'forest'. This sort of juxtaposition of seeming opposites is another feature that is fundamental to the composition of a classical garden. Alongside artificial/natural, Chinese gardens will purposely juxtapose light/dark, mountain/water, straight/curved, and large/small, among other similar binaries, to emphasize the notion of finding balance through tension. Similarly, the use of plain whitewashed walls sets off the more elaborate scenes depicted throughout the grounds. Chen points out that white is bland, but all colours come from it, and water is colourless, but all hues become richer when reflected in water.[80] In a Chinese garden, then, scenery is sought where there is no scenery, sound in soundlessness, and motion in stillness.[81]

There are two basic approaches to viewing a classical garden, which have been best described as 'in motion' and 'in position', motion and repose being another important binary often emphasized in landscape design.[82] The interaction between motion and repose is a key to garden design, one that Chen considers so

77. Ibid., 3.
78. Ayscough, 218.
79. Peng, 3.
80. Chen, 65.
81. Ibid., 65.
82. Ibid., 149.

important that he claims that if we simply grasp this dynamic, a garden's layout will resolve itself.[83]

In-position viewing refers to a lingering observation from fixed angles. Experiencing a garden in this way is ideal if the grounds contain numerous points of visual interest, and is generally preferred for smaller gardens.[84] The viewing points frame wider scenes than those seen through doorways and windows, and open onto the garden's best views.[85] In a garden designed with in-position viewing in mind, there will be numerous buildings that invite the visitor to linger.[86]

In-motion viewing is the opposite approach. The visitor tours the grounds, viewing each scene from changing angles. There are more and larger vistas available for viewing, and it is the preferred way of experiencing more expansive gardens.[87] The winding paths mean that even relatively static sights will seem to alter with the change in viewing angle.[88] Grounds designed for in-motion viewing will include many paths that wind around ponds, long corridors, and twisting walkways, and the view will change angles often.[89] In-motion viewing capitalizes on the way the path shifts the perspective of the visitor as she or he tours the grounds. 'Each move from one area to the next brings renewed scenes, extending the time and expanding the space for enjoyment.'[90]

The five gardens presented here, include Zuibaichi in Songjiang District, Qushui in Qingpu District, Guyi Yuan and

83. Ibid., 151.
84. Ibid., 15.
85. Lou 2003, 138.
86. Chen, 16.
87. Ibid., 15.
88. Lou 2003, 138.
89. Chen, 16.
90. Lou 2003, 61.

Qiuxiapu in Jiading District, and Yu Yuan in Shanghai's Old City—the only remaining samples of Shanghai's original classical gardens. Historically, Shanghai's gardens have been more prominent than many outsiders first realize, particularly those in Jiading and Songjiang. In fact, Nanxiang, a small town in Jiading and home to Guyi Yuan, was often favourably compared to Suzhou, with claims that its gardens rivalled those in the old centre of gardening culture in both beauty and number, despite Nanxiang being so small.[91] For over five hundred years, Suzhou held prime position in the gardening scene, standing with Yangzhou and Hangzhou as a representative of the Jiangnan landscaping design style. Yangzhou has long been known for its numerous gardens in which buildings were most prominent, while the water/mountain scenes occupied prime place in Hangzhou's gardens. Suzhou's style, known as *huicui* (薈萃, 'to gather important articles or personages', or in this instance, 'collage' or 'bricolage'), combined the best elements of all other styles of garden design. Shanghai's gardens built on the Suzhou tradition and, typical of much of Shanghai's culture, added a *haipai* (海派, 'fusion' or 'local colour') element.[92] Early interest in gardening blossomed in the city as people migrated from Suzhou to Shanghai beginning in the late Ming dynasty, resulting in hundreds of gardens emerging in Shanghai in the Ming-Qing period,[93] and all of the gardens explored in this monograph are examples of this period of growing interest in the landscaping tradition. The chapters that follow seek to open up for readers several avenues for how one might experience a garden, and so each chapter has

91. Yan Junming, *Enchanting Nanxiang* (Shanghai: Shanghai Cultural Arts Publishing, 2010), 5.
92. Tang, 10.
93. Ibid., 9.

been structured so as to allow a slightly different approach to the garden featured.

Extensive restoration works have been undertaken in all five of Shanghai's classical gardens, usually adhering to the garden chronicles that were written in the days of their construction. These chronicles usually did not contain images, but descriptions of the grounds and the philosophy behind their construction. Stanislaus Fung, an associate professor of architecture at the Chinese University of Hong Kong who has widely published in the field of Chinese garden history, explains, 'If we turn from twentieth-century writings to traditional writings on Chinese gardens, there is a significant "blackout". The use of visual means to record gardens takes a clear second place; the primacy of the word is conspicuous.'[94] The chronicles that describe the original gardens, then, are text-based and allow some space for interpretation when in the hands of a designer planning restoration works.

Like most of Jiangnan's gardens, Shanghai's five classical gardens have all been through a series of decline and rebuilding. Perhaps some will find it surprising that the Cultural Revolution (1966–76) was not as cruel to the region's gardens as to some other aspects of classical Chinese culture. Zhou Enlai, then premier of China, was instrumental in protecting the gardens from complete ransacking at the time, often referring to Jiangnan gardens as 'China's calling card', and urging that they be left intact for the time when China was ready to open its borders to the outside world.[95] Much damage had been done to the gardens prior to that time at the hands of the Japanese occupational forces, and that experience seemed to endow the gardens with a special measure of nationalistic signification that preserved them from being completely

94. Fung, 214.
95. Peng, foreword.

lost. It is hard to estimate how much the ransacking of the gardens during occupation, a matter that enraged local people much like the ransacking of Yuanmingyuan (the Summer Palace) in Beijing the previous century had been used to stir nationalist sentiment in the early 1900s, might have contributed towards a gentler treatment during the turbulent period of the Cultural Revolution. Also important is that, for many Chinese, the gardens represent nature at least as much as they represent culture, and so, while many of the inscriptions, paintings, and scrolls were destroyed, damaged, or lost during that period, the grounds remained relatively intact. The gardens of Suzhou were also among the earliest cultural sites to receive attention by rebuilding and restoration groups after the Cultural Revolution. The five classical gardens covered in this monograph were designated for preservation and restoration by the Nationalist government in May 1949, and so were under some measure of protection since that time.[96]

The restoration works that have since taken place are effective to varying degrees, with some very sensitive thought and planning being given to certain aspects of each garden. Chen Congzhou began setting out a systematic philosophy for garden restoration as early as 1978, shortly after the end of the Cultural Revolution.[97] Chen was Shanghai's most prominent figure in modern garden restoration and rebuilding, and also in designing contemporary landscapes. His writings will be cited often in this volume, particularly in the chapter on Guyi Yuan (Chapter 4), which he helped restore.[98] The work of Chen and other contemporary authorities on gardening culture can be seen as a part of a particular tradition (though not fully systematized) in the study of classical Chinese

96. Tang, 10.
97. Chen, 6.
98. Ibid., 6.

gardens—a system that has been underway for about seventy years and that combines Western and Chinese thought to create an understanding of classical garden design. Many of the Chinese scholars who have engaged in such work have gone about it with a conviction that Chinese culture of the past is valuable and can be preserved—even advanced—in the new, modern China.[99]

It should not be imagined that scholars giving attention to garden restoration after the Cultural Revolution were engaged in trailblazing work. Classical gardens have long been understood to be at the pinnacle of cultural expression in China, and invading forces were often known to ransack the grounds of both private and imperial gardens as a show of power.[100] Stanislaus Fung writes that 'the focus of narration in traditional writings does not shift from particularity toward generalization and abstraction, but broadens to show how gardens were part of the transformation of dynastic fortunes'.[101] Tales of rebuilding after such acts of aggression are common in the histories of many gardens in China generally, and Shanghai and the surrounding region specifically.

Even more frequently, restoration works were undertaken at different stages of each garden's life due to the more mundane problem of neglect. Chen Jiru (陳繼儒, 1558–1639), a renowned member of the literati during Ji Cheng's day (and apparently an acquaintance of Ji Cheng's associate Zheng Yuanxun), laments:

> I once said that there are four difficulties with gardens: it is difficult to have fine mountains and waters; it is difficult to have old trees; it is difficult to plan; and it is difficult to assign names. Then there are three easy things: the powerful can easily seize the garden;

99. Fung, 208.
100. Lou 2003, 145.
101. Fung, 215–16.

> in time, it can easily become unkempt; and with an uncultivated owner, it easily becomes vulgar.[102]

Many of the families that originally designed and built gardens in Suzhou and Shanghai were members of the *nouveau riche* during the Ming-Qing era, and the grounds were sometimes erected more as a show of wealth than with an understanding of the cultured tradition from which gardening first emerged. When the gardens passed hands from father to son, it was not uncommon for the care of the grounds to be neglected, or even for the whole compound to be sold or gambled away. But, keeping in mind the notion that it takes three centuries to properly construct a Chinese garden, such tales of neglect and restoration should not be surprising, as it follows a pattern of rise and decline that is common in the tale of families all over the world. In this and many other ways, each garden can be seen as an encapsulation of the unique stories of the women and men who designed, built, inhabited, maintained, and sometimes neglected it. In understanding the gardens of Shanghai, we may also gain a new appreciation of the story of the city and its people.

---

102. Quoted in Fung, 214.

2
Zuibaichi: A Historical Perspective

When looking back in time in search of Shanghai's earliest history, the eye naturally turns to Songjiang, in the southwest part of modern-day Shanghai Municipality. Long before the metropolitan sprawl on the Huangpu's banks burgeoned and became host to an influx of immigrants, both from overseas and from other parts of the Yangtze region, the city of Songjiang was the county and prefectural seat of Huating County and Songjiang Prefecture. Enjoying a history of more than nine hundred years, Songjiang is quite rightfully proud of the esteem that all things ancient can lay claim to in China. In keeping with this mindset, the classical garden situated in Songjiang, Zuibaichi (醉白池, or 'Drunken Bai's Pond'), claims to be the oldest of Shanghai's gardens, having first been founded by Zhu Zhichun (朱之純) in the ninth century during the Song dynasty and called Guyang Garden (谷陽園). Its history since then, however, has not been a continuous one, the original grounds having passed hands, fallen into disrepair, and been rebuilt numerous times. It was during the Ming dynasty that Jiangsu native Dong Qichang (董其昌, 1555–1636) rebuilt a garden on the same site, seeking to tap into the tradition stretching back hundreds of years and establishing a garden that has

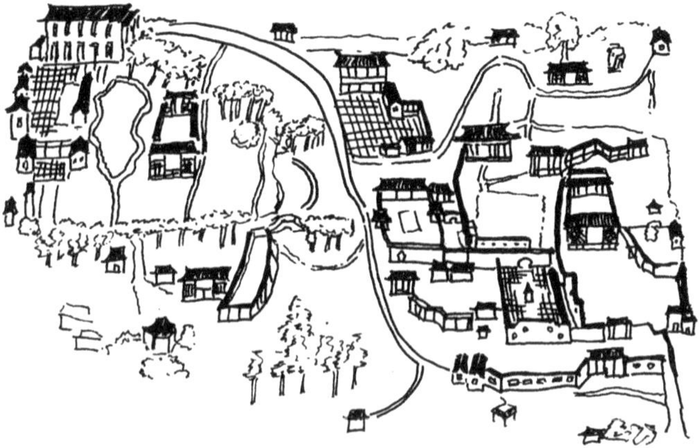

Figure 2.1
Layout of Zuibaichi (© 2015 Song Jie)

lasted into the present day. The garden was further expanded in the Kangxi period (1644–1722) of the Qing dynasty by Gu Dashen (顧大申), who renamed it Zuibaichi at that time.[1] The notion of following the traces that others have left behind is fundamental to Chinese garden construction,[2] and the history of the development of Zuibaichi from the Song through the Qing dynasty clearly demonstrates this value.

1. 《醉白池》，上海市地方誌辦公室，www.shtong.gov.cn/node2/node71994/node82435/node82441/index.html. Accessed on 29 October 2013. Hereafter cited as 'Zuibaichi'.
2. Peng, foreword.

Ming Dynasty: Dong Qichang

The owners of private gardens during the Ming dynasty were mainly retired court officials, landlords, scholars, and wealthy merchants. The etiquette system in imperial China placed restrictions on the extent of elaborateness and of the expenditure on, and build-up of, private property, preventing it from eclipsing anything the emperor had spent or built, so private gardens were kept small and secretive.[3] An emphasis on subtlety grew up to protect the garden owners from censure under the etiquette system, and much of Ji Cheng's advice in his handbook mentions keeping a garden low key and unobtrusive in its elegance. Dong Qichang, a painter, calligrapher, and possibly the most important of the wealthy men who retired to Jiangsu in the late Ming dynasty, was a student of Ji Cheng's patron Zheng Yuanxun.[4] That Dong's opinions on both painting and gardening were respected by other garden owners in the region is demonstrated by the fact that Zheng sought Dong's criticisms on some paintings and advice on the construction of his own famed grounds, the Ying Yuan (影園, 'Garden of Reflection') in Yangzhou, and even invited Dong to name the site.[5]

On the other hand, Dong and his family did not enjoy a good reputation among the common people. In 1604, there was a student uprising against him in the academy where he taught, forcing him into retirement.[6] Back in Songjiang, he and his sons were hated by

3. Cai, 122.
4. Ji, 11.
5. Zheng Yuanxun, 'A Personal Record of My Garden of Reflection', trans. Duncan Campbell, *Asian Studies Institute Translation Paper* (November 2004): 1–22. Print, 8–9.
6. Brandi Leigh, 'Dong Qichang', *The Art History Archive–Chinese Art*, 2008. http://www.arthistoryarchive.com/arthistory/asian/Dong-Qichang.html. Accessed on 29 October 2013.

the people of the city for their cruelty. On one occasion, his second son, Dong Zuchang (董祖常), beat a woman so severely that an irate mob turned out in opposition to the family, burning down their house. This led Dong Qichang to purchase a new plot of land in the western part of the city, the site of the old Guyang Garden.[7] He then rebuilt the garden there as a gathering spot for the literati of his day, setting it aside as a place for drinking wine and composing poetry with his friends. The reputation of the Dong family as cruel landowners was an amazingly enduring one, leading crowds to dig up and desecrate Dong Qichang's remains during the Cultural Revolution.

In the 1580s, Dong worked as a tutor for the family of Xiang Yuanbian (項元汴), one of the most prominent art collectors of the time. Later, he served as a government official in several capacities, including a stint as a tutor to the crown prince, then retired from 1605 through 1622. From 1622 to 1634, he was again in and out of office, according to the request (or whim) of the then-emperor, his former student.[8]

Hegel once described a classical Chinese garden as a painting that is both natural and poetic,[9] recognizing the connection between the work of a painter and that of a garden builder that has long been central to the Chinese approach to landscaping. Dong Qichang's own artistic achievements are reflected in the grounds of Zuibaichi, and it is representative of a true scholar's garden, with his system of thought made spatial.[10] Dong was, like many garden owners of the time, better known for his painting than his garden

7. Tang, 47.
8. 'Dong Qichang (1555–1640) and Art-Historical Painting'. http://www2.liu.edu/cwis/cwp/but06/hillwood/chinese/essays/no_essay_04.html. Accessed on 29 October 2013.
9. Cai, 123.
10. Tang, 45.

construction, but he was most renowned for his calligraphy. Though Dong's calligraphy was criticized during an early stage of the imperial examination, it later became his greatest strength.[11] The artistic style of his paintings was said to be influenced by calligraphy brushstrokes, though other aspects of his technique were often said to be wanting.[12]

Dong's aesthetic was largely influenced by revolutions that were going on in the arts during his lifetime, a 'revivalist' period that looked back to older traditions, especially those of the Yuan dynasty (1271–1368), and sought to revitalize them. By the middle of the Ming dynasty, some artists were looking back all the way to the Song dynasty for principles to be revived, re-examined, and reinvigorated, and Dong was one of the late Ming dynasty's most outspoken critics in this movement. He was especially active in the revitalization of landscape painting, which took centre stage during this artistic revolution.[13]

The Suzhou gardening tradition we know from later history really began during the Song dynasty. However, it was during the Ming dynasty that garden culture most flourished. During this period, scholars, poets, and painters gathered in Suzhou and Yangzhou. Some of the finest Ming dynasty cultural products came from these two cities, both of which were also home to numerous private gardens.[14] It was also during the Ming dynasty that a class of professional garden builders and designers arose and a theory of garden design was developed[15] and that Songjiang became a centre for *huicui* literati, drawing in the best ideas from the region.[16] Dong

11. Leigh.
12. 'Dong Qichang'.
13. Ibid.
14. Peng, 3.
15. Ibid., 3.
16. Tang, 45.

moved in the circles in which such theories were taking shape, even as he took part in a similar endeavour in revitalizing Song dynasty theories in painting. One can imagine that his interest in Song dynasty art would have made the site of the old Guyang Yuan, founded during the same period, particularly appealing to him. The Chinese approach to art sees beauty as rooted in the past, and there is almost always a generous measure of sensitivity to history in the Chinese aesthetic.[17] Dong was deeply immersed in this mindset, as is evident in his work in painting, calligraphy, and garden design.

Another important aspect of Dong's aesthetic was its roots in Chan Buddhism. A Jiangnan garden is meant to reflect the interests and tastes of its owner, making creative use of a small space to express the owner's personality[18] and Dong's commitment to Chan Buddhism influenced his approach not only to gardening, but to all art. He called his style of painting 'the southern school', in reference to his belief in enlightenment in art that emphasized introspection and self-cultivation, borrowing ideas that the southern school of Chan Buddhism applied to the spiritual growth of the individual. The southern school was known as the scholar's or literati school, and so held great appeal to a learned man like Dong. According to the southern school, there were two stages in a painter's development. The first involved being well-versed in tradition, and the second a moment of enlightenment. The northern school, by contrast, was more focused on meticulous technique, and therefore seemed to artists such as Dong to lack heart.[19] Zuibaichi likewise reflects this interest in both tradition and enlightenment in its repeated use of layering beautifully

17. Peng, 6.
18. Cai, 124.
19. 'Dong Qichang'.

crafted scenes with historical resonance in such a way that a new scene is revealed with an element of surprise as the visitor tours the grounds, demonstrating the collaboration between craftsmanship, tradition, and enlightenment that Dong believed to be at work in the arts.

A further influence of Chan Buddhism on Dong Qichang's aesthetic was his reliance on a master-apprentice system. Dong believed that each painter builds on the ideas of another, and that one's own work then serves as a model for later generations to build on and modify. Dong's efforts in building on another's work and leaving his own behind to be reworked by others were also evident in the development of Zuibaichi. His interest in gardening likewise allowed him to explore a tension between art and nature that was always one of his central concerns. He avoided naturalism in painting because he believed that art could not equal the beauty of nature. On the other hand, he was also convinced that nature alone could not achieve what an artist could do with the tools of his trade.[20] Every classical garden was constructed, at least in part, for the purpose of manipulating natural scenes in order to make them feel like living paintings, making landscape design an endeavour that would naturally appeal to Dong. The use of space and perspective in Zuibaichi reflects his fascination with these issues and their connection with the question of the relationship between art and nature.

Qing Dynasty: Gu Dashen

Gu Dashen was the next prominent figure who followed in Dong's footsteps in the cultivation of the site that had first housed a garden

20. Leigh.

in Songjiang during the Song dynasty. Gu, having inherited the garden, rebuilt and renamed it in 1650.[21] Being a poet himself, Gu renamed the garden 'Zuibaichi' (or, 'Drunken Bai's Pool') in honour of his own favourite poet, Bai Juyi (白居易), who had been instrumental in the early days of the development of private gardens in Luoyang during the Tang dynasty.[22] The 'drunken' modifier was partly in recognition of Dong Qichang's original purpose for rebuilding the grounds, and also in honour of a line from a poem that Su Dongpo (蘇東坡) wrote for Bai Juyi, 'In Memory of Drunken Bai'.[23] Gu was responsible for enlarging the grounds, and many of the features that he added to the site pay homage to Bai Juyi in some way, such as the Letian Hall, 'Letian' (樂天, 'happy-go-lucky') being a nickname Bai often employed for himself. The hall was first created in the early days of the garden's history during the Song dynasty and rebuilt many times over the years. With a 900-year history, it is the oldest hall in any of Shanghai's gardens.

The 700-square-metre lotus and water lily pond for which the garden is named is the most prominent feature in the grounds. It is not only the heart of the garden, but is also considered one of the most important historical sites in all of Songjiang.[24] Near the pond is a pavilion that bears a plaque with the inscription 'Zuibaichi' over its door, though the hall is officially named the Chishang Caotang (池上草堂), taken from Bai Juyi's poem 'Chi Shang Pian' (〈池上篇〉, 'On the Pond'). The poem reads:

21. Tang, 50.
22. Peng, 3.
23. Tang, 47.
24. Ibid., 48.

池上篇

十畝之宅，五畝之園。
有水一池，有竹千竿。
勿謂土狹，勿謂地偏。
足以容膝，足以息肩。
有堂有庭，有橋有船。
有書有酒，有歌有弦。
有叟在中，白須飄然。
識分知足，外無求焉。
如鳥擇木，姑務巢安。
如龜居坎，不知海寬。
靈鶴怪石，紫菱白蓮。
皆吾所好，盡在吾前。
時飲一杯，或吟一篇。
妻孥熙熙，雞犬閑閑。
優哉游哉，吾將終老乎其間。[25]

On the Pond

in a hut in the midst of the garden
surrounded by water and bamboo
do not comment on the thinness of the soil
for it is sufficient to bear our weight
here are halls and pavilions
                    bridges and boats
here are books and wine
                    music and song
the wisdom of old men fills the air
and we are content, asking for nothing
from outside the confines of these walls
the birds and the forestry shelter our silence
and like a turtle inside its own shell
I have no need to wonder at the vast seas beyond
the spirit of the crane fills the rocks and lotus

25. 白居易，《池上篇》，http://www.haoshici.com/Baijuyi46335.html. Accessed on 2 November 2013.

the whole scene lies before me
    and it is good
whether I raise my cup or chant a verse
with my wife and children at play nearby
the leisure of this home
    is all the pleasure I need

In later years, the tradition of scholars gathering in Zuibaichi continued steadily. By 1797, it had become a public gathering place for the people of Songjiang.[26]

The Restored Garden

All four sides of the Zuibaichi are bordered by some of the most significant features in the garden, including a Ming dynasty–style four-sided pavilion on the northeastern edge, where Dong Qichang liked to recite poetry and paint.[27] In later chronicles of the garden, Dong's intent that this hall be used for drinking and writing is often cited,[28] and the use of the grounds as a gathering place for poets and artists is firmly embedded in the minds of those who wrote about and visited the garden. Between this hall and the banks of the pond is a 300-year-old camphor, a tree that drops its leaves in the spring while simultaneously producing new ones.[29] Situated near the camphor is a 100-year-old peony tree that is said to be descended from a plant introduced into the grounds by Dong Qichang himself four hundred years ago.[30]

26. Tang, 50.
27. Ibid., 48.
28.《近代名園記述選錄》，上海市地方誌辦公室，www.shtong.gov.cn/node2/node2245/node69854/node69866/node69883/index.html. Accessed on 2 November 2013. Hereafter cited as 'Jindai Mingyuan'.
29. Valder, 275.
30. Tang, 48.

On the west side of the pond is a half-water/half-land pavilion. In summertime, the view of the sun rising over the lotus leaves from this pavilion is supposed to be one of the most spectacular sights in the garden.[31] The plaque that describes this pavilion includes the verse '小憩桐蔭坐曲廊／一規鏡檻綠泱泱／魚跳密澡深無影／燕掠平蕪靜有香', which means, 'At rest in the shade of the winding corridor, its balustrade green in the surrounding mirrored surface where the fish leap into the shadowy depths and the swallows sweep over the open spaces, a pleasurable tranquillity.' The pavilion is meant to be a 'scholar's hut', where the soothing scenes of the pond create the solitude necessary for literary and artistic pursuits.

Running southward from this small pavilion, skirting the western edge of the pond, is a corridor that houses thirty engravings of ninety-one important Ming dynasty celebrities from Songjiang.[32] The stones are all of equal size, 67 by 29 centimetres, featuring images of hometown heroes who were contemporaries of Dong Qichang, with Dong himself included among them. The men pictured here include mostly painters, poets, and scholars who were known as the pride of their generation. It is interesting to note that this collection of important personages also includes several prominent figures in the fields of agriculture, astronomy, and mathematics, symbolic of a paradigm shift in the late Ming dynasty that allowed people renowned in what we today call the sciences to be recognized alongside the more traditional *wenren* (文人), or literati.[33] The local hero Xia Yunyi (夏允彝) and his son Xia Wanchun (夏完淳) are likewise depicted on the wall. The pair are the subject of the author and poet Guo Moruo's 1943 historical

31. Tang, 48.
32. 'Zuibaichi'.
33. Tang, 54.

drama *Nan Guan Cao* (《南冠草》),[34] which retells the story of their resistance in the face of the invading Qing army during the late Ming dynasty and was used as a call to nationalism during the Japanese Occupation in World War II. Father and son are both buried at the nearby Xiao Kunshan Mountain at the Sheshan Resort.

South of the Zuibaichi and Letian Hall is the garden's original gate. Over 300 years old, the gate is representative of the Ming-Qing style, and the well-preserved wood carving above the doorway depicts scenes of scholarly life from that era. In order to get to the gate from the Letian Hall, visitors must pass through a small but stunning courtyard that highlights the technique of layering that is used to create a feeling of depth in the small confines of a Chinese garden. As one walks around the southern edge of the pond fronting the Letian Hall, the view through the doorway into this courtyard constantly shifts. A similar effect is achieved when the courtyard is crossed and the main gate approached, with the scene framed through the original main gate shifting from a moongate with a courtyard behind it to a beautifully laid-out miniature mountain scene. In the changing views and the use of layering, the visitor gains some insight into Dong Qichang's interests in the manipulation of space, perspectives, and perceptions of distance.

Inside the main gate, there are many intriguing features, beginning with a forest and mountain scene in the first courtyard, which is surrounded by a winding corridor that continues to play with the manipulation of space and perceptions of distance. One opening off this courtyard leads to a series of recently installed bonsai gardens, while the other leads to a courtyard with a large bust of Dong Qichang at its centre. Beyond this courtyard is a sort of inner garden, the eastern edge of which is a winding corridor punctuated

34. Tang, 53.

with windows of various shapes. The twists and window openings in the hallway, with plants strategically placed around them, is yet another variation on the themes of space and perspective. The layering of these elements and the constantly changing perspectives in this single corridor is one of Shanghai's best examples of this particular aspect of the Chinese gardening tradition.

The doorway that leads off this corridor brings the visitor back full circle, opening onto a courtyard that lies just north of Zuibaichi. This courtyard houses the Xuehai Tang (雪海堂, 'Plum Blossom Hall'), where the founder of the Chinese Republic Sun Yat-sen lectured on 26 December 1912.[35] Inside the hall is a display commemorating his visit. The inscription over the door of the Plum Blossom Hall is itself a cherished art piece, having been written by the Songjiang-born painter Zhu Kongyang (朱孔陽, 1892–1986) at the age of ninety-two.

With the Zhu inscription and the engravings of famous personalities, it is already clear that Zuibaichi is a treasure trove of cultural artefacts that are housed in some of the best preserved ancient buildings to be found in all of Shanghai, particularly those first constructed during Gu's rebuilding of the garden. This is exemplary of the *huicui* style in gardening for which Suzhou was famed, which implies a gathering of important people or cherished artefacts. This tradition flourished in Songjiang during the period of Zuibaichi's development, just as the city was becoming a renowned centre of learning and cultured study.[36] The engravings that remain on display are, therefore, considered highly treasured possessions. One of the oldest engravings in Shanghai today, a carefully preserved Ming dynasty piece, is housed in the Diaohua Ting (雕花廳, 'Flower Carving Hall'), located in the outer garden.

35. Tang, 57.
36. Tang, 45.

The *Shi Lu Jiu Huitou* (《十鹿九回頭》, 'Nine Deer Turning Back') engraving, situated at the northern edge of the old grounds, is another renowned piece. It depicts ten deer, nine of which are looking back over their shoulders in the direction from which they came while only one looks in the opposite direction. It is generally held that the implication of the image is that most people will always long for home no matter how far they might travel. The phrase 'nine deer turning back' became idiomatic in the region during the Qing dynasty much as the phrase *ye luo gui gen* (葉落歸根, 'fallen leaves return to their roots') is used to express a similar meaning today. There is some room for speculation about the meaning of this engraving, however, as only seven of the deer are actually facing backward in the same direction, and the one who is looking in the opposite direction is also looking over its own shoulder. It is possible that there are several layers of meaning attached to the piece—particularly as deer are often a symbol for Daoist immortals[37]—but the notion of longing for home is the one that has been the most enduring in the popular imagination. Nine deer mosaics can also be found in the floor tiles on the grounds of the park, all of which are looking back.

The Expanded Garden

Walking through the northern part of the grounds at Zuibaichi, the astute visitor will notice that there is less of the original garden visible in the layout of this section. A large stretch of open space is traversed by means of a path much straighter than one generally finds in a southern-style garden, lined by a straight row of trees that calls to mind the pattern that early twentieth-century Chinese writer and educator Ye Shengtao (葉聖陶) refers to as the 'military

37. Hardie, 103.

precision' more suitable to an imperial garden than to its Jiangnan counterpart.[38] The rockery and canal to the west of this path and the bridges and pavilions situated over the waters are rebuilt structures rather than restorations, but they provide a pleasant enough space for a stroll to the real highlight of the northern section of Zuibaichi. The pride of this rebuilt portion of the garden is the Flower Carving Hall, one of Songjiang's oldest buildings.[39] The Qing dynasty hall is home to two central gardens and four wings situated around the courtyard. The way the light falls across the grey roof tiles into the courtyard, playing on the rows of beautifully crafted windows, demonstrates the genius of the classical gardening style. As the day passes, the angle of light in the compound shifts, creating new patterns on the floors, walls, and carvings.

Once a self-contained garden owned by a citizen of Songjiang, the facade of the Flower Carving Hall is simple, with only minimal decorative adornments. Inside the wings at both ends of the courtyard are numerous intricate wood carvings of flowers. The images from the classical novel *Sanguo Yanyi* (《三國演義》, *The Romance of the Three Kingdoms*) visible on the roof beams, windows, and doors are a rarity in Shanghai. In the front and rear rooms of the hall are two large woodcarvings, *Baihua Qifang* (百花齊放, 'A Hundred Flowers in Bloom') in the front and *Chibi Dazhan* (赤壁大戰, 'The Great Battle of the Red Cliffs') in the room at the other end of the courtyard. 'A Hundred Flowers in Bloom' is remarkable not only for the intricacy of the floral image contained there, but its extensiveness, with no single flower repeated on the whole of the screen. 'The Great Battle of the Red Cliffs' represents a scene from shortly before the Three Kingdoms era in AD 208–9, a decisive battle fought on the banks of the Yangtze in the northwest part of

38. Ye Shengtao.
39. Tang, 50.

modern-day Hubei Province in which the united forces of Liu Bei (劉備) and Sun Quan (孫權) defeated the superior numbers of Cao Cao (曹操) and reunited the Eastern Han dynasty. Numerous accounts of this event can be found in various early histories, and *The Romance of the Three Kingdoms* includes an embellished version of the battle. The interior walls of the rooms surrounding the courtyard contain numerous additional smaller carvings. Many of the other original pieces in the Flower Carving Hall were destroyed during the Cultural Revolution. Those seen there today are what is left of what was once a magnificent collection.

On the outer edge to the north of the Flower Carving Hall, a pool has been constructed and surrounded with rocks and plants. This pond and its surroundings, along with the three newly completed courtyards containing an extensive bonsai collection, is among Zuibaichi's most successful constructions added as rebuilding and extension works were undertaken after the Cultural Revolution. Together, they help complete the picture sketched by the remnants of the Ming garden, many of which have been very skilfully restored.

The newly extended bonsai garden on the southern edge of Zuibaichi is perhaps the most impressive of recent restoration works in any of Shanghai's classical gardens. Though the bonsai is generally associated with Japan in the minds of most Westerners, it originated in China. In Chinese gardening, the term *penjing* (盆景) is more embracing than the equivalent of 'bonsai', *penzai* (盆栽), meaning 'potted plants'. *Penjing* refers to 'potted scenery' and includes potted plants, miniature rockeries, and occasionally water scenes. *Penjing* was practised for centuries in China—possibly up to two thousand years—before it was adopted in Japan.[40] In

40. Peter Valder, *The Garden Plants of China* (Portland: Timber Press, 1999), 31.

Zuibaichi, what was formerly a single courtyard has now been expanded into three interconnected areas, with a moongate between the two newest courtyards framing views of a canal around which the various *penjing* have been arranged. The potted scenes in these courtyards include both plants and rockeries, with some of the mountain scenes being particularly good examples of the lifelikeness that the form intends to capture. Pines, the tree most often used in *penjing*, appear in most of the tree arrangements at Zuibaichi, with several plantings of cedars and cypresses also on display.

Zuibaichi is perhaps the most accessible of Shanghai's classical gardens for the visitor who is less initiated in the traditions from which these sites emerged. The layering effect of the original garden is well executed and prominently displayed. It is easy to see how these techniques have been employed to expand the sense of space in what is actually a very small compound. The playfulness that lies at the heart of the Chinese approach to gardening is foregrounded, and it is not unusual to hear guests express surprise when they come across some particular feature there.

This accessibility would likely be a matter of pride to both Dong Qichang and Bai Juyi, were they able to see it today. Part of Dong's agenda in his own art was to restore simplicity and vitality in painting, reacting to what he saw as decadence in the landscape painting of his time.[41] Bai Juyi went about his own work with a similar goal in mind. Known for a clear, direct, accessible style of poetry, Bai was said to rewrite any part of a poem that his servants could not understand. It is fitting that the garden that bears his name is one of the simplest and most direct expressions of classical gardening techniques that Shanghai has to offer.

41. 'Dong Qichang'.

3
Qushui Yuan: Space and Text

Established in 1745 during the Qianlong period (1736–95) of the Qing dynasty, Qushui Yuan (曲水園, 'Garden of Meandering Waters'), in the city's Qingpu District, served as a resting place for guests in the Qingpu City God Temple compound from its earliest years, rather than as a part of a single family's residential estate. The overall layout of the garden centres around its religious function, and many of the inscriptions are laden with spiritual significance. The temple to which the garden was originally attached was founded in the Ming dynasty in 1573. Like Yu Yuan in the Old City of Shanghai, the garden's proximity to the temple made it a central gathering space for the community, and a shopping bazaar has long stood in the area around the garden. In the early days of the Chinese landscaping tradition, temple gardens and private gardens were very closely linked, and only later were they sometimes thought of as two separate traditions, though the distinction was never thought to be as clear as that between private and imperial gardens.[1]

1. Peng, 1.

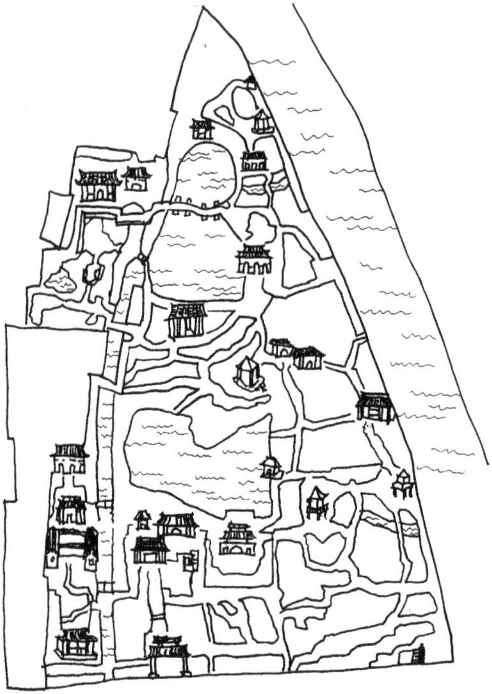

Figure 3.1
Layout of Qushui Yuan (© 2015 Song Jie)

Most cities and even small towns in ancient China had a city god temple (城隍廟, *cheng huang miao*). The city god was often a prominent local personality who, after his death, was venerated through ancestor worship to a point that he was taken as a protector of the whole city. All the residents of the town would worship at his shrine, seeking his favour and providence. The city god was believed to be the last judge whom a soul would face on earth before passing on to the afterlife. Qushui Yuan is one of the four classical gardens in Shanghai that was historically connected to its town's City God Temple. Since undergoing extensive restoration works in 2006, Qingpu's Chenghuang Miao, neighbouring Qushui Yuan, has housed the Qingpu Museum.

Bordered on the southern and eastern edges by the city moat and using it in many 'borrowed views', Qushui Yuan has been an important feature of the ancient city since its earliest days. The funds for constructing the original site were donated by local residents, with each person in the city paying a levy of one *wen* (文), or one penny, per year, so the garden was originally known as One-Penny Garden (一文園, Yiwen Yuan). Over the next decade, the grounds were improved and expanded. In 1784, Wang Xiyi (王希伊), then governor of Qingpu, built a rockery on the east side of the Deyue Xuan (得月軒, 'Gallery for Catching the Moon'), added a few small bridges, and created several new scenes within the compound. The garden's shape at this point in its development is what is now meant when reference is made to its 'original form'—though in fact construction on the garden continued for twenty more years after this—and it was at this time that it became known as Ling Yuan (靈園), or the 'Garden of the Soul'. The shape, and even name, of the garden changed often over the years, with new scenes being added when care of the grounds passed hands. Changes of this sort happened in every garden with the passing of time, such changes being inevitable in an endeavour that was understood to require a timeframe spanning several generations. Modifications of this sort generally occurred more frequently in the temple gardens than in privately owned sites, since the transition from one caretaker to the next was less likely to be smoothly conducted from father to son.

An important date to note in the development of the garden is 1798, when Liu Yunfang (劉雲房), a scholar and poet from the neighbouring Jiangsu Province, renamed it Qushui Yuan. The name comes from the Eastern Jin dynasty (317–420) calligrapher Wang Xizhi's (王羲之) preface to *A Collection of Poems Composed at the Orchid Pavilion* (《蘭亭集序》, *Lan ting ji xu*), in which he reports a famous event that took place in a garden in Shaoxing. According to the story:

> In AD 353, one year before his official retirement, Wang invited forty individuals to pass the Purification Rites festival with him at a famous pavilion in the Guiji area (modern-day Zhejiang Province) where he was serving as governor. A poetry contest was held alongside a stream, down which cups of wine were floated; anyone of the forty-two in attendance who could not finish his composition by the time the cup arrived would have to pay the forfeit and drink. This gathering soon achieved legendary status, and references to it occur throughout the poetry and painting of later eras.[2]

Liu Yunfang's 1798 visit to Qingpu at the invitation of the county governor was a significant event for the whole city, and a gathering of local scholars turned out at the temple garden to meet with Liu, all of them writing verses to commemorate the occasion. The gathering and poetry that grew out of it brought to mind Wang's famous party years earlier, inspiring Liu to rename the grounds Qushui Yuan in honour of that connection. The term *qushui* is often found in Chinese gardens, and special tables with miniature waterways built in to replicate the game that occupied

2. 'Six Dynasties Calligraphy', http://depts.washington.edu/chinaciv/callig/7calsixd.htm. Accessed on 3 November 2013.

Wang and his companions are sometimes included as part of a garden's furnishings. In describing the elements of a garden Ji Cheng mentions *qushui*,[3] referring to it as 'an almost essential component in any cultured person's garden'.[4] In Qingpu's Qushui Yuan, the name is meant to point to the gathering of scholars, and also to the waterways around which the grounds have been laid out, particularly the canals dredged to bring water from the town lake into the garden.

Qushui Yuan was destroyed during the Taiping Rebellion in 1860. It underwent its first rebuilding from 1884 through 1910, with numerous scenes being added at this time. In 1911, it was made a public park, and eighteen years later, local businessman Zhang Jingzhou (張景周) made a large contribution to fund further restoration works.[5] The grounds were renamed Zhongshan Park (not to be confused with present-day Zhongshan Park in Changning District) during the 1927–29 restoration works. During World War II, the garden was again destroyed. What remains in the grounds today was mostly rebuilt after 1949, with the garden being set aside as a historic site under the protection of Qingpu County in 1959.[6] In 1980, the site was rechristened Qushui Yuan, with the local government investing over RMB800,000 from 1983 through 1986 to rebuild the whole site according to the original records of the Qing dynasty garden's construction.

The grounds of Qushui Yuan are divided into four major parts. The southern portion is dedicated to the three halls, with

3. Ji, 110–11.
4. Ibid., 133, note 156.
5. 《曲水園》，上海市地方誌辦公室，www.shtong.gov.cn/node2/node71994/node82435/node82440/index.html. Accessed on 29 October 2013. Hereafter cited as 'Qushui Yuan'.
6. 'Qingpu Museum.' http://museum.shqp.gov.cn/gb/content/2009-02/23/content_237430.htm. Accessed on 3 November 2013.

Ninghe Hall (凝和軒) playing the central role in the scene. The western edge of the garden is dedicated to several important buildings and corridors, many of which house paintings or calligraphy. The eastern section is dedicated more to plants, and the central and northern areas of the garden are focused on the two lotus ponds. All of the pavilions in the garden follow the original rise and fall of the land on which the garden was established, and each pavilion is topped by a symbol of some sort, either an animal, flower, or gourd.

The garden was originally built around twenty-four scenes. Most have been rebuilt today, with several more added to other parts of the grounds. Four of the original twenty-four scenes—the Baiyun Wu, the Mibai Pavilion, the Qing Xu Jing Stage, and the Huanbi Hall—are no longer found inside the grounds, not having been rebuilt in recent reconstruction efforts.

The Eastern Garden

Typical of a temple garden, all of the scenes in Qushui Yuan surround a central pavilion, the Ninghe Hall, originally built in 1767. Two other important structures stand on either side of this hall, the Youjue Hall (有覺軒) on the west and the Huashen Hall (花神軒) on the east, a linear, symmetrical arrangement of the buildings that is rare in a Jiangnan garden. The linear layout is 'hidden', though, by the use of walls that screen each of the pavilions from the others. Access from one hall to the next is only offered by way of winding paths, masking the symmetrical arrangement that is so untypical of a Jiangnan Garden.

In 'The Twenty-four Scenes of Ling Yuan' (〈靈園二十四咏〉, *Ling yuan er shi si yong*), a Qing dynasty poem describing the views contained in the original garden, Jin Erxi (金二溪) writes of the Ninghe Hall:

> 凝和堂
>
> 粵惟靈園成，斯堂實經始。時和民亦和，萬人樂
> 輸此。四序靄春光，熙熙介蕃祉。
>
> Hall of the Coalescence of Harmony
>
> upon entering Ling Yuan
> the tour only really begins here
> where all the people love to tell
> of their harmonious life together
> a melody arranged like the misty spring morn
> as the sunlight settles between lush branches

Directly behind the Ninghe Hall is a lotus pond, with a miniature mountain of huang stones lying on the opposite bank separating this lake from the garden's other major waterway, a second lotus pond. The lotus, a flower that appears in almost every Chinese garden, is loaded with symbolic significance in Chinese thought. Buddhist teaching associates it with purity, while in secular thought, the fact that its flower, pod, and bud all appear at the same time has made the lotus a symbol for the life cycle, and its abundance of seeds connects it to abundant progeny. The Chinese names for the flower, *lian* (蓮) and *he* (荷) are homonyms for 'continuous' (聯) and 'peace' (和), making for another auspicious connection.[7]

As the name Qushui suggests, water is the major element around which the garden's scenes have been constructed. To the west of the Ninghe Hall, a straight canal runs from the southern edge of the garden to the northern lotus pond, connecting all of the

7. Valder, 227–30.

major waterways in the grounds to the city moat surrounding the walls of the compound. The canal partitions off a small courtyard in the southwestern corner of the garden, where some of Qushui's most intriguing features are found. The garden's marble boat (丹居非水, Dan Ju Fei Shui) is fronted by the canal on the east, and on the west is a tiny koi pond, with the Xiyanghong Banlou (夕日紅半樓, 'Half-Tower of the Red Sunset') perched atop the rockery that faces the marble boat on the other side. The koi pond appears quaint but unremarkable most of the time, but at midday when the sun is at its height, especially on a summer afternoon, the pool gains a striking clarity, with scenes of the boat, rockery, and pavilion reflected in its waters in the most vivid colours, making it appear that the fish swim in and out of the windows of the buildings. As with many of the features in this garden, though, the view is only at its most captivating for a short time each day, and only in particular weather conditions. This characteristic of Qushui Yuan drives home a commonly observed point about Chinese landscaping—the notion that a garden can only be really experienced by a visitor who is willing to tour its grounds at various times of the day and of the year so that she may get a feel for the changes of light, season, and weather.[8]

In a courtyard to the north of the marble boat, fish pond, and rockery, separated by an elevated corridor, is the four-sided Youjue Hall, with a century-old osmanthus tree beside it. This hall, first built in 1745 and rebuilt in 1887, is one of Shanghai's most significant examples of ancient architecture, noted for its beamless ceiling, a type of structure that is thought of as the height of garden architecture. Today, the Youjue Hall is one of two such buildings remaining in Shanghai. Sitting behind the Youjue Hall is the Yushu

8. Lou 2003, 143.

Lou (禦書樓, 'The Royal Library'), which houses several examples of ancient inscriptions, a Confucius shrine, and an intriguing collection of paintings.

The marble boat and the Deyue Hall separate this courtyard from the Ninghe Hall and the rest of the garden. The Deyue Hall and the marble boat's upper floor house some of the key artefacts in the garden. The second floor of the marble boat served as a study for the local poet Shen Shoudong (沈瘦東, 1888–1970) in the years leading up to the Cultural Revolution. Samples of Shen's work, copies of his books, and furnishings from his study occupy the hall now.

Connected to the marble boat and forming a long corridor alongside the southern lotus pond is the Deyue Hall. The windows of this hall on both the upper and lower floors open onto the pond, framing some of the garden's most exquisite views. The opposite wall displays calligraphy and paintings, creating a contrast between the painted and natural scenes that line the hall on either side. Jin's poem describing the twenty-four scenes in Qushui says of the Deyue Hall:

> 得月軒
>
> 樓台瞑色深，此間月先到。畫壁素光流，縱橫披
> 荇藻。道人證空明，夜夜來相照。
>
> Gallery for Catching the Moon
>
> beside the gallery in the night's deep hues
> the moon arises through the darkness
> its light flows along whitewashed walls
> over the open blossoms in the pond
> the priest opens up the space for our minds
> as the moon nightly ushers in the dawn

Beneath the Deyue Hall, separating it from the lotus pond, is the north–south running straight canal that connects to the city moat, skirted by a causeway with a humpbacked bridge. The waters of this canal reflect off the whitewashed walls and along the underside of the roof tiles, creating shifting patterns with the play of light on the water. Though unfortunately the grounds are not open to guests at night when the gallery's function of 'catching the moon' would be best experienced, this clever use of water and light mimics the effect, to some degree. This manipulation of light, along with the borrowed views that utilize the city moat in other parts of the grounds, is an example of pulling elements from outside the walls into the garden, of viewing 'gardens outside of gardens' and 'scenery outside of scenery'.[9] The tradition of 'borrowing views' (接景, *jiejing*) is a technique that makes a conscious attempt to pull elements from outside the immediate scene in to form part of the effect, creating greater depth and scope than a very confined space might otherwise offer. The seven different approaches to borrowing scenes are borrowing from afar, borrowing from nearby, mutual borrowing, borrowing from above, borrowing from below, and seasonal borrowing.[10] Qushui Yuan makes effective use of all seven techniques, with the seasonal borrowing offering some of the subtlest views in the garden. Professor Chen Congzhou gives high praises to such borrowings of light, shadows, and fragrances.[11]

The Central/Northern Garden

Perched on the north, east, and south banks of the southern lotus pond are three pavilions, Xiao Hao Liang (小濠梁, 'Moat-Bridge

9. Chen, 34.
10. Cai, 126.
11. Chen, 34–35.

Pavilion'), Yingxi Ting (迎曦亭, 'Pavilion for Greeting the Dawn'), and Huang Dui Feilai (恍對飛來, 'Communing with the Immortals'). Each of these pavilions become focal points for views framed by the windows of the Deyue Hall, even as they serve as points from which to view the pond, framing scenes from different angles. Jin's description of Xiao Hao Liang, the northernmost of the three pavilions, reads:

> 小濠梁
>
> 秋水讀南華，所思在濠濮。清景無時無，惠莊不可作。會心豈在遠，我亦知魚樂。
>
> Moat-Bridge Pavilion
>
> when autumn waters are heard in the south
> thoughts turn to the moat
> the lasting clarity of this view
> beyond what even the poets can express
> then understanding bridges the gap,
> revealing to me the fishes' pleasure

The pavilion on the pond's eastern bank is the Yingxi Ting. In Jin's verse, its function is contrasted to the Deyue Hall, which stands on the opposite side of the pond from this pavilion. The names of these two structures similarly highlight this contrast, reflecting the sense of poetry underlying the design of the garden. Jin says of the Yingxi Ting:

> 迎曦亭
>
> 紅日升東隅，孤亭翼然跂。滿園金碧輝，都在朝陽裡。譬若月之得，亦如雨之喜。
>
> Pavilion for Greeting the Dawn
>
> when the sun rises in the east
> uplifting the wings of the pavilion

the garden is filled with lights of gold and green
besieged by the reign of the sun's light
as fitting for us as for the moon
and as pleasing as the falling rain

The pavilion on the water's southern bank sits alongside the few Taihu stones in this section of the garden, where it is huang rockeries that are more prominently employed to create harder edges to contrast with the large expanses of water. The Taihu stones on the southern bank of the pond and several of the carvings on the rocks nearby carry a more spiritual, contemplative significance, with the carvings representing the Eight Immortals and the virtues they embody. The pavilion situated here, called the Huang Dui Feilai, similarly carries the religious connotations that are sprinkled throughout this garden. Jin writes of this pavilion:

恍對飛來

臨流築小榭，泉石相回環。一望碧無際，魚戲蓮
葉間。日夕坐忘返，為愛隔湖山。

Communing with the Immortals

overlooking the flow from the pavilion
that faces the turns of the huang hills
viewing the endless verdure
where fish play amongst lotus leaves
as the sun sets we forget to return home
lost in longing for the distant landscape

These three pavilions together are meant to represent heavenly, earthly, and human happiness, in line with Daoist thought. The cluster of scenes created by this pond and the three pavilions on its banks emphasize the garden's original role as part of the temple complex, highlighting the spiritual nature of the atmosphere one expects to find in a Jiangnan garden, as is reflected in Ji Cheng's admonition, 'Never say there are no Immortals on earth, for this is a fairy-land in the world of men.'[12]

The huang stone arrangement on the northern bank of the pond, facing Huang Dui Feilai, separates this lake from the other large body of water contained inside the garden's walls, a second lotus pond. The choice of huang stones for the formation is important, as they better represent the solidity that contrasts with water, providing the sort of philosophical symmetry that Chinese art aims at, the same aesthetic central to Chinese landscape painting. Ji Cheng writes of huang stones, as compared to Taihu, that they are fairly regular in shape with comparatively flat surfaces and so are good for stacking together.[13] While Taihu stones can sometimes seem to have a flowing effect, huang stones are harder and more solid. This attribute is clearly observed in the central mountain that divides Qushui Garden's north and south sections.

Atop this hill sits the Jiu Feng Yi Lan (九峰一覽, 'Nine Peaks with One View'), a pavilion that offers an excellent overview of both lotus ponds. The use of different types of trees to optimize the changing views with the change of seasons is well executed, with pines, maples, osmanthus, and magnolias visible through the windows on each side. Lying to the north of the central rockery, the second lotus pond dominates the view in that direction. Like its counterpart to the south, buildings are incorporated into the scenes on the three sides of the pond. The building to the north is the Qing Lai Shan Fang (清籟山房, 'Chamber of the Hillside's Clear Sound'). This chamber is central in the view from the porch of the Zhu Xie (竹榭, 'Bamboo Gazebo') on the bank at the foot of the huang rockery, with a nine-turn bridge visible in the foreground. The bridge and the chamber are, along with the Youjue

12. Ji, 107.
13. Ibid., 135, note 188.

Hall, among the garden's most remarkable architectural features. Of the Qing Lai Shan Fang, Jin writes:

> 清籟山房
>
> 丹室靜無塵，一徑入香積。松陰篩鶴糧，真修於焉息。清風時一來，萬籟與俱寂。
>
> Chamber of the Hillside's Clear Sound
>
> the refinery and kitchen of the monks
> calm in the absence of worldly pursuits
> solitude seasons the holy men's diet[14]
> breathing fresh life into their seclusion
> with the coming of the clear breeze
> myriad soft sounds complete the quietude

On the eastern bank of the pond is the Jing Xin Lu (鏡心廬, 'Mirror Cottage'). Like the koi pond next to the marble boat in the southwestern courtyard of the garden, this hall is best viewed at a particular time of the day. However, unlike that pond, the Jing Xin Lu is not unremarkable at other times. Rather, it is slightly disconcerting, a large white space that threatens to overpower the soothing effect of the pond. In late afternoon, however, the purpose of this blank space is revealed. Known as a mirror wall, its whitewashed surface hovers over the pond, creating a screen where the reflections of the early evening sun on the water are displayed. It is a stunning visual effect, perfectly blending with the more subtle features of the lotus pond, and befitting its name. It is commemorated in Jin's poem among the famous scenes in the original garden:

---

14. My rendering, 'holy men's diet', comes from an understanding of the original 鶴糧 that can be found at http://www.zdic.net/c/4/3c/91706.htm.

鏡心廬

玲瓏九曲橋，窈窕三間屋。倒影入鏡中，照見鬚眉綠。誰攜並州刀，剪取吳淞幅。

Mirror Cottage

a delicately wrought winding bridge
ending at a three-roomed hut
their reversed image in the mirror,
the green pond's unshaven face
—who has carried this severed piece here
this swathe cut from the docks of the Wusong?

Ji Cheng's general description of mirror walls gives a more technical view of how the effect was created during the Ming dynasty and before. In a section on whitewashed walls, he writes,

> Traditionally walls have been whitewashed with a mixture of paper-pulp and lime. Those who like a fashionable look and want a glossy finish cover the wall with white wax and polish it up. Nowadays, people use yellow sand from rivers and lakes, mix it with good quality lime and put on an undercoat, then they put on a top coat of more lime, brushing it lightly with a hemp brush, which makes it naturally so bright you can see your reflection in it. If it gets dirty, the dirt can be washed off at once. This is known as a 'mirror wall'.[15]

Chen Congzhou writes, 'A garden without water, clouds, shadows, sounds, morning twilight, and sunset is a garden devoid of natural beauty. For these, though ethereal, set off the actual scenes in the garden.'[16] This concept is demonstrated often in the buildings of Qushui, where light and shadow play together on the

15. Ibid., 96.
16. Chen, 150.

surfaces of the buildings and the waters they surround, creating a great variety of scenes throughout the day. In order to fully appreciate this, Qushui Yuan, perhaps more than any other in Shanghai, demands multiple visits over a long period of time, and at different times of the day and year.

The Eastern Garden

The eastern edge of the garden opens onto the city moat, with a covered corridor, the Lubo Lang (綠波廊, 'Green Waves Corridor') forming the outer wall on this side of the grounds and drawing the moat in to form part of the view available in the garden's scenes. The area between this corridor and the two ponds, along with the rest of the southeastern section of the grounds, is dedicated mostly to plants. Several Taihu stones are scattered among the plants, creating a multitude of miniature scenes within this portion of the grounds. There are three spots where stones are built into the walls and many more where they are installed up against a wall. Turning again to Ji Cheng, we find a description of the technique at work here. He writes, 'What are known as precipitous mountains are built up against walls, so that the whitewash surface acts as paper and the rocks as the painting upon it.'[17]

The main plants used in the grounds of Qushui Yuan are the osmanthus tree and the bamboo, with an extensive bamboo grove occupying the south-easternmost corner of the grounds. The bamboo is the plant most often associated with China, and is a prominent part of Chinese art and technology. With its pliable, resilient form, it has come to symbolize 'gentlemanly qualities, lasting friendship, longevity, and hardy old age'.[18] The osmanthus

17. Ji, 107.
18. Valder, 90.

is a fragrant tree whose aroma permeates the air in the autumn, making it a favourite garden plant because it makes its presence felt without calling overt attention to itself. Its name *gui* (桂) is a homophone for 'noble' (貴), and the plant is often used to symbolize this trait.[19]

Near the bamboo grove is a peony garden. The tree peony, unlike the osmanthus, is a showy, extravagant flower. Blooming in the spring, the peony was originally admired for its medicinal purposes, but the striking pinks and reds of its blooms inevitably led to its (now more common) ornamental use in Chinese gardens.[20] The peony is regarded as 'the king of flowers' in Chinese tradition, and stands for wealth, honour, and the aristocracy, but also for love and affection.[21]

Just south of the peony garden, leading back towards the Ninghe Hall and the main gate, are a pair of courtyards with the inscriptions *qing mei jing song* (青梅勁松, 'fresh plum blossoms and strong pines') and *ying ge yan wu* (鶯歌燕舞, 'singing orioles and dancing swallows') hung over their respective gates. These courtyards are meant to represent winter and spring, respectively, and the plants each houses function together with the rock formations to create the atmosphere appropriate to these seasons.

Within the grounds of Qushui Yuan, there are over fifty trees that are more than a century old. Many of them are marked with plaques stating their age.

19. Ibid., 313.
20. Ibid., 170.
21. Ibid., 174.

The Western Garden

No garden is considered complete until it is adorned with distinguished words and texts,[22] often in the form of inscriptions. In Qushui Yuan, the western portion of the garden is dedicated to this purpose. In the northwest corner is a courtyard devoted mainly to the display of calligraphy. A long corridor wraps around the courtyard, boasting numerous engravings on its walls, and engraved stones are also scattered among the plants and pavilions of this courtyard along the banks of a narrow stream and koi pond. Three important blue stone steles stand among the calligraphy engravings in this portion of the garden. The stele of *Poetry on a Visit to the Garden by Zhao Yixi* (趙宜喜) is 28 centimetres high, 83 centimetres wide, and engraved with a running regular script, a calligraphic style that first appeared in the third century and was considered fully matured by the seventh century. It has four poems composed by Zhao Yixi, a Qing dynasty magistrate of Songjiang, Liu Yunfang, and two other scholars. *Poetry on a Visit to the Garden by Wang Quan* (王勸) is also displayed in the northwestern courtyard. Standing 28 centimetres in height and 82 centimetres in width, the stele was inscribed in regular script, recording poems composed by Wang Quan and other scholars from the region. The inscription of *Poetry on the Garden by Zhu Delin* (祝德鱗) is likewise in running regular script, engraved by Zhu Delin and including poetry by Zhu and Liu. All these steles were carved and installed in the garden in 1789, marking the gathering of scholars

22. Lu Shaoming, 'From Syntax to Plot: The Spatial Language of a Chinese Garden', University College London, Bartlett School of Architecture. http://www.sss7.org/Proceedings/03%20Spatial%20Analysis%20and%20Architectural%20Theory/067_Lu.pdf. Accessed on 6 November 2013. 067: 9.

that convened to welcome Liu to Qingpu.[23] It was customary in ancient days for a prominent visitor to the garden to write commemorative verse to leave in the grounds in this way.[24]

Today, Qushui Yuan continues to have a very local flavour, befitting its original role as a garden in the City God Temple complex. Local residents, young and old, are often seen enjoying the grounds, though few tourists visit the site. The absence of tourism is partly because this is the least accessible of Shanghai's classical gardens by public transportation, and perhaps partly because it is less accessible intellectually to one who does not already have a good grasp of traditional Chinese gardening culture. The subtleties of some of the installations, the abundance of calligraphy, and the patience required to catch a view of many of its features at their best make it less attractive for the general tourist, even as it appeals to scholars or other guests with a more discerning eye and more experienced in the nuances of gardening culture. It is exemplary of the secrecy of a Chinese garden, which Scottish-born expert on China and its gardens Maggie Keswick points to in her foreword to Ji Cheng's classic text, saying that this system of walls-within-walls and courtyards-within-courtyards only reveals itself to one who really enters into the work itself.[25]

The approach that the current local government in Qingpu has taken to restoration and preservation of Qushui Yuan and the City God Temple has proven to be quite successful. The first district in Shanghai to give serious, sensitive consideration to restoration even as it re-envisioned contemporary uses for these ancient sites, Qingpu set the standard for similar works to take place in other parts of the city over the decade that followed its

23. 'Qingpu Museum'.

24. Lu, 067: 9.

25. Ji, 15.

recommissioning of the City God Temple as a museum. Jiading District modelled some of its more elaborate plans—including the Confucius Temple, the Nanxiang Temple, Nanxiang Old Town, the Fahua Pagoda, and the two classical gardens in Jiading District—on the restoration and rebuilding works that were so successfully executed in Qingpu. Qiuxiapu and Guyi Yuan, the subjects of the two chapters that follow, were two of the main beneficiaries of the work done at Qushui Yuan.

4

Guyi Yuan: Revival, Restoration, and Expansion

During the prosperous period of the Ming dynasty, the town of Nanxiang in modern-day Shanghai's northwestern district of Jiading, was one of the wealthiest and most cultured cities in the region. The Tang dynasty text *San Yi Ji* (《三易集》, 'Three Dialogues') includes 〈南翔八老人詩序〉, or 'Preface to the Eight Scholars of Nanxiang', which tells of the town's well-known members of the literati. In a region renowned for its wealth, art, and scholarship, it is no surprise that many gardens were constructed during the heyday of private gardening culture. Indeed, the fact that two gardens still stand in Jiading District today, Guyi Yuan (古猗園) and Qiuxiapu (秋霞圃), is evidence of the stature the area once enjoyed.[1]

The earliest gardens in Nanxiang were built during the Song dynasty. With Hangzhou serving as the capital during the Southern Song (1127–1279), the Jiangnan area became known for its gardens at this time. It was during the Ming dynasty, though, that gardening culture really exploded in the region, and that Nanxiang became well-known for its scholar's gardens. The famed Ming-era

1. Tang, 27.

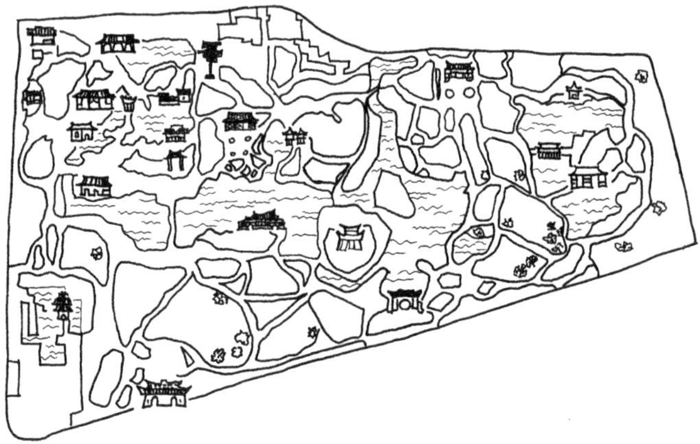

Figure 4.1
Layout of Guyi Yuan (© 2015 Song Jie)

gardens of Nanxiang were known for their simplicity and elegance, and for their natural appearance. Eleven gardens were built in the town during the Ming dynasty, with another seventeen being added during the Qing period.[2] According to the signage inside the grounds of the garden, the old local saying is that 'Suzhou gardens are famous around the world, and Guyi Garden is Shanghai's best.'

Song Dynasty Roots

Nanxiang takes great pride in being one of the oldest parts of Shanghai, and Guyi Yuan is one of the main locales where the ancient flavour of the town is foregrounded. As in all classical gardens, the notion of a connection with history is important in

2. Yan, 5.

Guyi Yuan, and this tie to the ancient is particularly highlighted in specific features in the grounds, such as the pair of cypresses in front of the Meihua Ting (梅花廳, or 'Plum Blossom Pavilion') that are thought to be over a thousand years old.[3] In fact, the character *gu* (古), meaning 'ancient', was added to the garden's name when it was renovated during the Qianlong era (1736–95) more than two hundred years ago by the site's owner at the time. An understanding of the history in which the garden is steeped is one of the best methods for gaining some appreciation of the grounds.

Nanxiang records indicate that Lu Hong (陸紘) first built a garden on the site where Guyi Yuan now stands during the Song dynasty from 1225 to 1227, planting bamboo and flowers.[4] It is impossible to say exactly when Yi Yuan was established, but the earliest date that we can be sure of from the extant records show that the garden, as we know it, began during the Ming dynasty between 1522 and 1566.[5] While most of the features in the garden date from the Ming dynasty and later, there are several reminders that the site has a much longer history than that. The Usnisa Vijaya Dharani sutra pillars are the oldest tablet inscriptions in Nanxiang.[6] There are three such sutra pillars still in existence in Shanghai, the other being in Songjiang.[7] The pillars, one dating from 865 and the other from 875, are nearly 10 metres high. Previously, they were located in the Nanxiang Temple's main hall. During the Qing dynasty, one sutra pillar was destroyed in a typhoon and reconstructed in 1796. When the temple was destroyed, one pillar fell, but the other remained intact. Both were moved to Guyi Yuan in 1959, when the

3. Ibid., 45.
4. Ibid., 150.
5. Tang, 28.
6. Yan, 62.
7. Ibid., 63.

garden was enlarged. One was placed in front of Weiyin Ge (微音閣) and the other in front of Nan Ting (南廳, 'South Pavilion'). In 1969, the sutra pillar in front of the South Pavilion was devastated by lightning, and only one-third of it now remains. During the Cultural Revolution, the Buddha on top of the other stele was cut off. The renovations that took place in the garden from 1980 to 1982 included getting both sutra pillars listed as cultural relics under the protection of the Jiading District government.[8]

Nan Ting, which accompanies the stone pillars, is a later structure, having been built in the Ming dynasty and served as the study of the official Li Yizhi (李宜之) when he owned the garden.[9] Scenes from the four seasons are displayed in the courtyard around it, and a Qing dynasty couplet by Ye Changchun (葉長春), author of four volumes of poetry, was inscribed in 1986 and installed in the pavilion, reading:

> 月來滿地水
> 雲起一天山
>
> moonlight pools on the earth
> cloudy peaks rise to the heavens

Since the Ming dynasty, Nan Ting stood at the foot of Zhuzhi Shan (竹枝山, 'Bamboo Hill') to the southwest.[10] The Qing dynasty *Chronicle of Yi Garden* (《古猗園記》, *Gu yi yuan ji*) cites the conventional wisdom in gardening circles which states that creating favourable geographic conditions for a garden largely depends on having hills in the grounds (踞一園之形勝者莫如山),[11] so Zhuzhi

8. Ibid., 63.
9. Tang, 44.
10. Yan, 15.
11. Quoted in 'Guyi Yuan', 《古猗園》，上海市地方誌辦公室，www.shtong.gov.cn/node2/node71994/node82435/node82439/index.html. Accessed on 29 October 2013.

Shan is obviously one of the key reasons such an effective layout has been achieved in Guyi Yuan.

Another of the garden's noted ancient sites, the Putong Pagoda (菩同塔), dates from the Song dynasty.[12] Over seven hundred years old, this pagoda was previously located in the Lotus Pond of the Nanxiang Temple. The six-sided stone tower, more than 2 metres high, is made of carved stone. The inscriptions on the pagoda were carved in 1222. It was moved to Guyi Yuan during the 1959 renovations of the garden, and in 1960 was also listed as a key cultural relic under the protection of the local government.[13]

Ming Dynasty Heyday

Though Guyi Yuan has roots that extend much further back in time, it is in the Ming dynasty that the real story of the garden begins. Built by an assistant prefect of Henan Province, Jiading native Min Shiji (閔士籍), and designed by the famed Nanxiang bamboo carver Zhu Sansong (朱三松), the garden was reputed for being graceful, elegant, and patriotic.[14] However, according to local legend, the grace and elegance of the grounds in its early days made its patriotism suspect. It was said that the earliest owner of the garden had returned to Nanxiang after fulfilling his official duties and constructed the site when he heard that it was his mother's dream to walk through an imperial garden. Though his efforts pleased his mother, the emperor was unhappy when he heard news of it. When it was reported that the grounds contained a four-sided pavilion, the emperor took this as a symbol of the garden owner's desire to become king of the south, and had the whole family

12. Yan, 64.
13. Ibid., 65.
14. Ibid., 10.

executed.[15] An alternate story says that the emperor took offence at a pair of scholar trees that were planted in the courtyard of the Yiye Tang (逸野堂), a symbol only allowed in imperial gardens. When the emperor sent his officials to investigate, the garden's owners had just cut down one of the trees, and so the family was spared the severe punishment that would have surely followed. One of the scholar trees is said to still stand in the Yiye courtyard today, a second tree having been planted in later years to replace its lost partner.[16]

Perhaps it is partly due to such legends of imperial dissatisfaction that Guyi Yuan has a clear thread of political engagement woven into the fabric of its story, as is evident in various features that were added to the grounds in later years. Many of the gardens were used for political purposes at various times in their histories, partly because they are secretive spaces, and some of the names of features in Guyi Yuan reflect various political movements,[17] with the Weiyin Ge (微音閣, 'Small Voice Gallery') and Quejiao Ting's (缺角亭, 'Lost Corner Pavilion') overt anti-Japanese connections being some of the latest examples of the political tone of the garden's features.

Besides Nan Ting and Zhuzhi Shan, many other features in the compound date from the garden's golden age in the Ming dynasty, and when restoration works were undertaken in 1985–87, the new scenic area was based on Ming dynasty plans.[18]

The Buxi Zhou (不系舟, 'Anchorless Boat'), another of Guyi Yuan's Ming dynasty instalments, is reputedly the best marble

15. Ibid., 8.
16. Tang, 37.
17. 'Guyi Yuan'.
18. Yan, 11.

boat in Shanghai.[19] Its name comes from a line of Bai Juyi's poetry. These marble boats were common in classical gardens, being ideal sites for entertaining guests, and many scholars from Suzhou often came to sit in Guyi Yuan's marble boat to drink, write poetry, and paint.[20] This Qing dynasty verse by Liao Shoufeng (廖壽豐) describes the marble boat:

> 不系舟
>
> 十分清水雙簷影
> 百葉蓮花萬里香
>
> The Anchorless Boat
>
> two eaves shadowed in depths of green
> blend with the surrounding fragrance of lotus

According to local legend, by the time Guyi Yuan had been made the City God Temple during the Qing dynasty, the Anchorless Boat housed a dining hall where patrons ate *xiaolongbao* (小籠包) dumplings, and meaty *mantou* (饅頭) steamed buns, both specialties of Nanxiang. Years later, a Nanxiang native opened a store selling the two delicacies on Shanghai's Central Xizang Road, the beginning of a long tradition of exporting the famous Nanxiang *xiaolongbao*.[21]

The Jiu Qu Qiao (九曲橋, 'Nine-Turn Bridge') stretches across the Lotus Pond south of the Buxi Zhou, the entryway to the northwest part of the garden where many Ming dynasty features are situated. Hehua Chi (荷花池, 'Lotus Pond') and Yuanyang Hu (鴛鴦湖, 'Mandarin Duck Lake') both contain numerous blooms, featuring various types of lotus, including red, white, yellow,

19. Ibid., 17.
20. Tang, 38.
21. Ibid., 39.

compound colour, single lobe, and double lobe.[22] Local residents have a long tradition of visiting the garden in the summer to see the lotuses in bloom, and an annual lotus expo attracts many guests from the region each year. Numerous contemporary paintings of the lotus remain continuously on display in the winding corridors of Guyi Yuan. The flower is admired not only for its brilliant blooms, but also for the expanses of green its huge round leaves create, its pleasing fragrance, and even the sound of the rain falling on its leaves.[23]

Major waterways in Guyi Yuan include the Xi'e Pond (戲鵝池, 'Goose Playing Pond'), the Yuanyang Lake, the Lotus Pond, and the Guishan Lake (龜山湖, Turtle Lake).[24] During the Ming dynasty, the Xi'e Pond became the central feature of the garden,[25] and today, its surroundings include the Baihe Ting (白鶴亭, 'White Crane Pavilion'), Buxi Zhou, Fuyun Pavilion (浮雲閣), Zhuzhi Shan, and the Lost Corner Pavilion.[26] The use of water has always been one of the main reasons that Guyi Yuan is seen as such an excellent example of the Jiangnan-style gardens, with water helping to create scenes within scenes through reflections, and also juxtaposing movements above and below the surface of the water.[27]

White cranes have been brought in to occupy the Xi'e Pond. Along with the Baihe Pavilion, these birds commemorate the story of the founding of Nanxiang, which is likewise mirrored in the town's name (南翔, or 'Southward Flight').[28] According to the legend, nearly two thousand years ago, a pair of white cranes were

22. Yan, 41.
23. Valder, 231.
24. Tang, 32.
25. Yan, 29.
26. Tang, 37.
27. Ibid., 33.
28. Yan, 17.

in the habit of perching on a rock in a local farmer's field. When a monk happened to walk by and heard of this, he built a temple on the site. Generations later, a Tang dynasty monk visited the site, and two cranes followed him on his journey to the temple. The man who owned the field around the temple area decided to donate his land to the temple, and when he did, it was discovered that the rock where the cranes liked to rest was inscribed with a poem (presumably left by the birds) saying that they had flown south and would no longer return.[29] This legend has been the source for many of the names of structures in Guyi Yuan.[30]

Also located in this area of the garden is the oldest structure remaining in the grounds, Yiye Tang.[31] Though Yiye Tang is another example of Ming dynasty architecture, its furnishings are Qing-style.[32] This verse by the late Qing poet Zhang Senyu (張森于) is displayed in the Yiye Tang, describing the view from the hall:

> 古木蔥籠飛鳥止
> 漪漣蕩漾任魚游
>
> above, birds stop in the verdant canopy
> below, fish swim through ceaseless ripples

The name Yiye Tang indicates that this is a secluded spot intended for reflection. It is considered the most important of Guyi Yuan's buildings. It is a four-sided pavilion, offering an excellent overview of the scenery inside the garden. It once contained a Dong Qichang calligraphy that read 華岩墨海 (*hua yan mo hai*, 'magnificent mountains and inky sea'). According to the stories told at the time, upon hearing of the garden's beauty, Dong visited

29. Ibid., 101.
30. Tang, 39.
31. Ibid., 36.
32. 'Guyi Yuan'.

Guyi Yuan and met with other scholars in the Yiye Tang to discuss Buddhist scriptures. When he opened the window on the west side of the hall, he saw the rockeries on the banks of the Xi'e Pond, which appeared dark in the setting sun. When he saw the view, he composed this calligraphy to leave in the grounds to commemorate the gathering. Today, a piece by contemporary calligrapher Tang Yun (唐雲) stands in its place.[33]

It was also during the Ming dynasty that the garden was first named Yi Yuan when it was re-established by Min Shiji.[34] The name 'Yi Yuan', inspired by the *Book of Songs* and *Odes to Qin* (琴賦) refers to the artistic conception of beautiful, green bamboo. Bamboo has been a central part of Guyi Yuan since the Ming dynasty, with many of its features centring around this plant, including Zhuzhi Shan, where a hillock with a small grove of various types of bamboo forms a pleasing scene.[35] According to the garden annals, this is meant to be the most graceful scenery in the garden. Water and bamboo have been seen as the two main features of the garden for over five hundred years,[36] with bamboo forming about one-third of the features in the grounds of Guyi Yuan.[37]

That the use of bamboo is a central feature of the grounds is perhaps not surprising when it is remembered that the designer employed by Min Shiji was Zhu Sansong, a famous bamboo carver from Jiading.[38] Bamboo was a common feature in Nanxiang Gardens as early as the Song dynasty, and the plant was an important part of Jiading's development and culture in ancient times, as

33. Tang, 36–37.
34. Yan, 10.
35. Ibid., 26.
36. Tang, 31.
37. Ibid., 27.
38. Yan, 10.

well as a part of Nanxiang's origin as a township.[39] Being abundant in Jiading, bamboo was used for construction, as a food source, and in the arts. The Jiading Bamboo Carving Museum, near Qiuxiapu, showcases some pieces of the bamboo carving tradition for which the area was famed, including pieces carved by Zhu. The bamboo groves in Guyi Yuan were the garden's claim to fame in the early days, being known not only for its wide variety of species of bamboo—numbering over fifty—but also for its distribution of the plants within the grounds. Among the species featured in Guyi Yuan today are fernleaf, hedge bamboo, black bamboo, and various examples from the Pleioblastus and Phyllostachys families.[40] In the garden's bamboo groves, views of the trees themselves are obviously part of the charm, but considered equally beautiful are the shadows they cast on the ground, water, and whitewashed walls of the buildings that surround them. Each year, a bamboo culture festival is held in Guyi Yuan, which includes displays of bamboo crafts.

From a cultural perspective, the importance of bamboo extends well beyond the crafts made from it. It has often featured in literature since the Tang dynasty and is perhaps the most frequently depicted plant in Chinese art. It is the plant that is most readily recognized in early Chinese art and, during the Tang dynasty, it was a central feature in the increasingly popular bird-and-flower style of painting. It is also the subject of the oldest extant botanical monograph in China, dating from AD 460.[41] Known as one of the 'three friends of winter' (歲寒三友, *sui han san you*), alongside the pine and the plum blossom,[42] bamboo is admired for its ability

39. Ibid., 42.
40. Ibid., 152.
41. Valder, 90.
42. Ji, 133, note 153.

to thrive in any weather conditions. Its reputation for resilience and hardiness is furthered by the fact that the hollow bough of the tree allows it to bend without breaking. This admirable quality of the plant was used throughout history, especially at the collapse of the Ming dynasty, as a metaphor to both describe and encourage southern scholars and politicians who refused to serve under non-Han masters from the north.

The bamboo-like resolve of the people of Jiading was tested at the end of the Ming dynasty during the Jiading San Tu (嘉定三屠, 'Three Massacres of Jiading') in 1645, when nearly all of the local residents were killed by invading Qing armies because they had refused to conform to the Qing government's strictures concerning hair and dressing styles.[43] One of the victims of the cruel massacre, Tang Yu (唐預), wrote this poem about Yi Yuan:

猗園詩

台榭參差處，林塘散影清。雀喧鶯默坐，魚駭鶴
潛行。獨樹偏成趣，群花不識名。鄴候書滿架，
靜對綠窗明。

A Poem of Yi Yuan

stages and pavilions in irregular arrangement
trees' crisp reflections fill the pools
orioles sit quietly amid the sparrows' clamour
and the fish are startled as cranes wade in the shallows
a single tree leans across the waters
surrounded by blossoms that have forgotten all the
names
the scholar's shelves stacked with books
quietly faces the moonlight outside the window

43. Tang, 41.

After six different periods of expansion, Guyi Yuan was a very well-developed garden by the end of the Ming dynasty. It was said that there were over twenty different views in the grounds before the Qing dynasty commenced.[44]

Qing Dynasty Transition

During the winter of 1746, Guyi Yuan was bought by Ye Jin (葉錦), a scholar from Dongting Shan.[45] Ye came from a very wealthy Suzhou family,[46] and it was under his oversight that the garden was extensively revamped, a project that took two years to complete.[47] During the time the Ye family owned the garden, Shen Yuanlu (沈元祿) wrote the chronicle of Guyi Yuan (《古猗園記》). The chronicle includes a description of the location of the garden, being on the west side of the Guangfu Temple and with its south gate facing a creek. It also described the features inside the walls, including rockeries, corridors, mountains, buildings, waterways, and the marble boat, which had all been built in the Ming style.

In 1789, when the expense of caring for the garden threatened to exceed what one family could shoulder, the citizens of Nanxiang collected funds to purchase the garden, transformed it into the City God Temple, and donated it to the state. At this point, it became a place of worship and recreation for the community, and a teahouse and restaurant were added to the grounds.[48] In 1806, the citizens of Nanxiang again collected funds to rebuild and repair the grounds.

44. Ibid., 29.
45. Ibid., 29.
46. 'Guyi Yuan'.
47. Yan, 10.
48. Tang, 29.

The Meihua Ting dates from the Qing dynasty and still retains a Qing style in its timber structure and upturned eaves.[49] The pavilion, made completely of wood, with black posts and violet windows, stands in the oldest part of the garden. Covered with delicate engravings following a plum motif and topped with simple cornices, the Meihua Pavilion courtyard is an excellent example of how a sub-garden can be built within a larger complex to explore a specific sub-theme of a particular site's overarching theme or narrative, as described in Lu Shaoming's insightful discussion of Yu Yuan.[50] The plum blossom garden that surrounds the pair of ancient cypress trees in front of the hall completes the thematic study with samples of red and green plums, making the area particularly fragrant when the flowers are in bloom.[51] North of the Meihua Pavilion is the Plum Blossom Stone Tablet Gallery, home to numerous calligraphy pieces further extolling the beauties of the plum blossom. Second only to bamboo in symbolic significance, the plum blossom is tightly entwined with Chinese culture, appearing often in paintings and literary works. The plum blossom flowers in winter, and its tree is gnarled and crooked, leading it to 'embody the virtues of purity and ability to endure hardship, evoking . . . the doughty Confucian hero devoted to principles come what may'.[52]

During the Qing dynasty, many canals were dredged to extend the waterways around the garden's central waterway, Xi'e Pond.[53] On the banks of these canals, arbours and buildings were constructed to echo the rivers that wound through the grounds.

---

49. Yan, 19.
50. Lu, 067: 9–12.
51. Tang, 34–35.
52. Valder, 106.
53. Yan, 29.

Besides the Meihua Pavilion, other buildings that were added at this time include the Yiye Pavilion, Cui'ai Lou (翠靄樓, 'Tower of Green Mists'), Liudai Xuan (柳帶軒, 'Willow Belt Hall'), and Chunzao Tang (春藻堂, 'Spring Waters Hall'). All were constructed in an elegant style typical of Qing dynasty architecture, serving both as sites from which to view the scenery, and as part of the scenery to be viewed from alternate angles. A winding corridor connects the various buildings in this part of the garden.[54] The Qing dynasty chronicle of the garden states that the Yicui Ting (怡翠亭, 'Emerald Harmony Pavilion') once stood nearby amid the bamboo,[55] but it is missing in today's garden. A pavilion of the same name stands in the new extensions to the grounds in the northeast corner.

The use of peonies to brighten the grounds of Guyi Yuan also dates back to the Qing dynasty. North of the Quxiang Lang (曲香廊, 'Corridor of Winding Fragrances') is a super-double violet peony bush that was planted during the reign of Tongzhi (1862–74). The double variety of the peony was first mentioned in Chinese literature in the eighth century. Those who cultivated the earliest known samples of the double peony began a fad that resulted in exorbitant prices being paid for the newest varieties. These large flowers are often called 'hundred-petalled' or 'thousand-petalled'.[56] The flowers in the Quxiang Corridor can be as large as a bowl, and the plant has produced a record number of sixty-two blossoms at one time. The much younger plant on the south side of the gallery, only right around a century old (just after the end of the Qing dynasty), produces blooms on higher branches than its northern

54. Ibid., 31.
55. 'Guyi Yuan'.
56. Valder, 176.

counterpart, and its flowers have petals that overlap. Both plants were donated to the garden.[57]

A poem by Zhu Wei (朱瑋), an early/mid-Qing poet and painter, describes the evening view at Guyi Yuan:

> 猗園晚眺
>
> 平沙邐迤水雲鄉，桐竹蕭森坐晚涼。遠近青光添淡沲，勾人詩興是斜陽。
>
> The Evening View from Yi Yuan
>
> clouds wind through the canals flowing over the countryside
> as we sit in the secluded bamboo grove in the cool evening
> all around the soft slanting rays intensify the pale light
> the setting sun engendering a seductive poetic impulse

Guyi Yuan did not escape the political scene of the Qing dynasty any more easily than it did that of earlier or later times. In June 1853, the Small Swords Rebellion (a local group proclaiming allegiance to the Taiping rebellion) held secret meetings at Xiao Songgang (小松崗, 'Little Pine Hill'), located in the Xi'e Pond.[58]

Turbulence and Restoration

In the modern era, Guyi Yuan was caught up in some of the most turbulent political conflicts seen in Shanghai. After the commencement of the January 28 Incident (sometimes known as the Shanghai Incident) of 1932, Japanese occupational forces seized Nanxiang. During this incident, Guyi Yuan was occupied for two months, during which time it was badly damaged, including the

57. Yan, 39.
58. Tang, 42–43.

loss of several scrolls of Dong Qichang's calligraphy.[59] In May 1933, the citizens of Nanxiang donated funds to partially repair the garden, and also to build the Quejiao Pavilion, or the 'Lost Corner Pavilion'. The pavilion was first built as a symbol of the 'national humiliation' suffered at the hands of the Japanese occupational forces during the Mukden Incident of 1931, the event that presaged the Japanese annexation of Manchuria. With its northeast corner missing, the building symbolizes the three provinces in northeast China that were occupied by Japan and renamed Manchukuo. The pavilion is situated at the highest point in the garden (Tang, 40).

Built atop Zhuzhi Shan, the Quejiao Ting is a square pavilion, but its roof only has three upturned eaves, each shaped like a raised fist. Upon hearing news of the Japanese annexation of the northeastern provinces, the people of Nanxiang were enraged, and so erected this potent symbol of protest. The sixty residents who helped fund the construction of the pavilion were led by three prominent citizens who were particularly outspoken against the Japanese invasion, Zhu Shouming (朱壽明), Chen Shaoyun (陳少芸), and Wang Xuan (王萱).[60]

The numbers 8–1–3 have a significance to the people of Shanghai in the same way that 9–1–1 sets off traumatic memories for many Americans today. On 13 August 1937, the Battle of Shanghai, the first of twenty-two engagements between the Nationalist army and Japanese invading forces, was commenced—engagements that eventually saw Shanghai fall to the Japanese invaders. Nanxiang was one of the towns in Shanghai that provided a large number of volunteers for the Nationalist armies, and it suffered badly during the invasion of Shanghai. Knowing that the Japanese soldiers who entered Nanxiang would not be pleased to see a structure with

59. 'Guyi Yuan'.
60. Yan, 76.

such an overt message of protest standing in the garden of the City God Temple, the citizens were naturally afraid that the Quejiao Pavilion would be destroyed. Hoping to prevent this outcome, the townspeople reportedly hung an inscription inside the pavilion that read, 'China can never be humiliated. Japanese pirates, don't be savage. This pavilion is blessed by God. Your Mikado will be killed by heaven.'[61] The next day, when the invaders took possession of the garden, they were terrified when lightning struck just as they tried to enter the Quejiao Pavilion, making the clenched fists on the eaves seem as if they were ready to strike. The soldiers were said to have retreated in fear at the sight.[62]

After China's war against Japanese aggression, a cultural association, the Wei Yin She (微音社, 'Small Voice Society'), was formed by the alumni of the Nanxiang Public School. The society constructed the Weiyin Gallery and proposed extensive renovations of the whole war-ravaged garden. The 10-metre tall pavilion stands to the northeast of the Quejiao Ting.[63] The Weiyin Gallery is located to the north of the stone sutras, and is meant to echo the elegance of the pillars.[64] Its inscription, along with that over the garden's main gate, was written by the renowned calligrapher Hu Juewen (胡厥文). In 1946, local residents Chen Shaoyun, Fang Jian'ge (方劍閣), and Zhu Suwu (朱蘇吾) repaired the Quejiao Pavilion, the Baihe Pavilion, and Nan Ting. Other wealthy merchants from Nanxiang contributed funds to repair the marble boat and add more plants to the grounds. After 1949, several rounds of restoration works were undertaken at Guyi Yuan, including

61. 中華不可侮，倭寇勿猖狂。本亭有神佑，天滅爾天皇。Translated by Yan Junming.
62. Yan, 180.
63. Ibid., 78–79.
64. Ibid., 19.

an expansion of the entire garden in 1958, followed by a second expansion in 1963.[65]

During the Cultural Revolution, many of the buildings in Guyi Yuan were damaged, including the loss of the Buddha from the top of one of the Tang dynasty sutra pillars. However, many scholars continued to take an interest in classical gardens even through the social upheaval—often covertly—and were able to put restoration works into effect as early as 1973. The efforts gradually increased to the point that a full rebuilding of the garden was undertaken in 1977, just after the close of the Cultural Revolution. Emphasis during this rebuilding was on the construction of the rockeries, the waterfall, canals, and ponds. The Putong Tower was also moved into the Lotus Pond at this time.[66]

Further renovations took place from 1981 onward, much of it being overseen by Professor Chen Congzhou of Tongji University, with the final stages that included the eastward expansion of the grounds after Chen's death being completed in 2009 for the sixtieth anniversary of the People's Republic of China. Chen, a leading voice in the restoration of classical gardens in recent decades, was one of the scholars who maintained his interest in Chinese gardens throughout the Cultural Revolution, an interest that had first grown out of his studies of *kunqu*, or Suzhou opera. Much of his study focused on the intersection of gardening with other art forms.[67] Sometimes the best way to achieve the intended effect in a garden, according to Chen, is through poetic annotations or inscriptions,[68] and his contribution to Guyi Yuan includes an inscription of one of his own poems in the Meihua Pavilion.[69]

65. Ibid., 11.
66. Ibid., 11.
67. Chen, 6.
68. Ibid., 83.
69. Yan, 33.

One of the key texts that Chen cites in considering how gardens should be rebuilt is Tao Yuanming's *Taohua Yuan Ji* (《桃花源記》, 'Land at the End of the Peach Grove'). Though gardening is never mentioned in the essay, Chen points out that all of the principles of garden culture can be seen in it.[70] Familiarity with Chinese literature and art is an important part of constructing a garden, whether it be a specific work that is quoted in the grounds, or a piece with a more general intertextual connection, such as Tao's essay. Chen writes, 'It is indeed true that, without being thoroughly impregnated with traditional Chinese culture, it is impossible to master the essential[s] of the art of garden designing.'[71] Chen's own immersion in Chinese art and literature has played an important role in maintaining Guyi Yuan's connection with classical literature and other art forms. This sort of work Chen calls true restoration, as opposed to rebuilding, because it is more firmly rooted in the texts to which the original grounds were tied.[72] Chen does not, however, disparage the notion of reusing the remnants of an older garden for a rebuilding project, 'mak[ing] use of what is left and redesign[ing] the garden as we see fit'.[73] There are times when a project of this sort might be a more fitting use of the remains of an old garden rather than attempting to restore the grounds. The Tan Garden (檀園), which has been rebuilt on its original site just north of the temple in Nanxiang, is an example of a work that Chen would have termed a 'rebuilding', and one which would have likely received praise from him.

In 1985, an extension was added to the southeast part of Guyi Yuan, with the newly added Qingqing Yuan (青清園, 'Verdant

70. Chen, 83.
71. Ibid., 88–89.
72. Ibid., 90.
73. Ibid., 92.

Garden') forming a garden within a garden. Based on paintings by Qing dynasty scholar Zheng Banqiao (鄭板橋), the bamboo garden brought in more than thirty different breeds of bamboo to fit in with the overall theme of Guyi Yuan. There is nothing else quite like Qingqing Yuan for variety and concentration in any other Jiangnan garden.[74] With all of the expansions that have been made to Guyi Yuan, it is now the biggest of Shanghai's gardens, spanning ten hectares.

Some of the most recent expansions at Guyi Yuan seem to need a little more time to mature, but much of the work has been thoughtfully laid out. Most of the older parts of the garden are in the northwestern portion of the compound. Upon entering from the main gate, the residential buildings from the old garden lie to the west. In the original layout of Yi Yuan, about half of the land was set aside for residential buildings, and the other half for the scenery, a typical configuration in a classical garden.[75] With the garden's expansion and the decline of the residential buildings over the years, these structures now form a very small part of the grounds overall.

From the main entrance to Guyi Yuan, the Yuanyang Lake sits straight to the north on the other side of a grassy area, with a path winding around it. Along the way are an extensive rockery, some century-old peonies, and a regular retinue of exhibits that alternate to suit the season. The Nine-Turn Bridge crosses the pond, leading to Nan Ting and the Quejiao Pavilion on the other side of the water. Following the path towards Nan Ting leads in a circle around all the significant older buildings, the Yuanfei Yuyue Hall (鳶飛魚躍軒, 'Hall for Viewing Bird and Fish'), the Yiye Hall, and the Baihe Pavilion. At this point, the path arrives at the Xi'e Pond

74. Tang, 33.
75. Ibid., 34.

and the rest of the older structures, including the marble boat and the Meihua Pavilion.

Moving eastward from the Meihua Pavilion leads to the expanded sections of the garden, which cover roughly half the grounds. The path allows for a pleasant walk winding through art galleries, well-crafted twisting corridors with windows of various shapes, bamboo groves, and numerous newly built structures. Each path in the garden should be noted for the motif laid into it, which is often used to give a hint about the feature that lies at its opposite end. Two of the more interesting new installations in the expanded grounds are the bonsai garden and the extensive rockery neighbouring it. These two features of the newer portions of Guyi Yuan demonstrate Chen's notion that successful rebuilding efforts can complement the restoration of an older garden in magnificent fashion.

Chen, who was well acquainted with the difficulties of restoration and rebuilding, believed that it required a skilled craftsman to conceptualize a garden from old plans. When one's skill or the records left behind were insufficient, he (perhaps surprisingly) advocated leaving the garden in disrepair rather than furthering the mess that it was in. The main belief that underpinned his work and his teaching on garden restoration, repair, and rebuilding, was that the study of old gardens and their history was a valuable work, and that the gardens were worth restoring for modern enjoyment and appreciation. Aware of the uphill battle this represented at the time Chen was working, he reminded, 'It is not easy to keep a garden in good repair. It is even less so to renovate it.'[76]

76. Chen, 116.

5
Qiuxiapu: A Literary Perspective

Though Yu Yuan is by far Shanghai's best-known garden, Qiuxiapu (秋霞圃) rivals both its storied past and its elegance of design. Yu Yuan's main appeal for tourists—its location in downtown Shanghai—has made it the one garden in the city that both foreign and domestic tourists are likely to know today, but throughout history, Qiuxiapu has been at least as renowned, and its name shows up in more literary references than any of the other four classical gardens in contemporary Shanghai, meaning that there were many volumes of writings to be sorted through when the garden was restored.[1] Its grounds are likewise overflowing with literary references in the place names, inscriptions, and calligraphies displayed there, making it the best example in the city of the intertextual character of a Jiangnan garden.

1. Wu Yi, 吳義，《秋霞圃志（重修本）》編纂小記，www.shtong.gov.cn/node2/node70393/node70403/node81552/node81554/userobject1ai105092.html. Accessed on 29 October 2013.

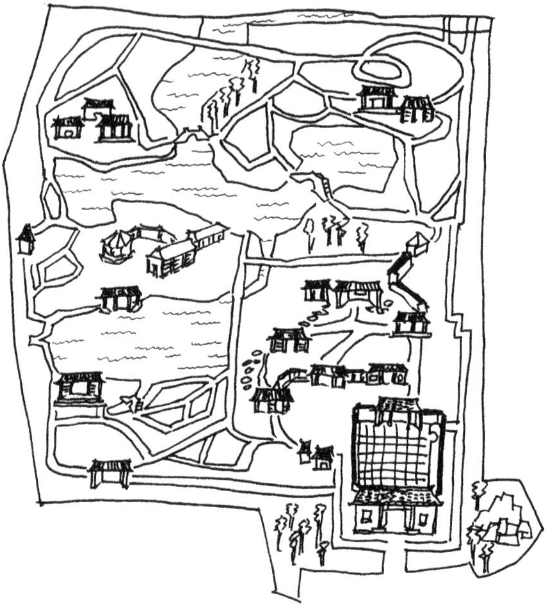

Figure 5.1
Layout of Qiuxiapu (© 2015 Song Jie)

History of Development

From its earliest days, Qiuxiapu was an 'urban garden', built in the bustle of Jiading for private enjoyment. Though most of the structures inside the grounds today are examples of Qing-style architecture, Qiuxiapu is a Ming garden, with connections reaching as far back as the Southern Song dynasty (1127–1279).

What we today know as Qiuxiapu began as four separate gardens, those of the Gong (龔) clan, the Jin (金) clan, the Shen

(沈) clan, and the City God Temple (城隍廟). It is the latter that dates from the Song dynasty, and since it has continuously remained a part of the grounds where the current garden sits, Qiuxiapu claims to be Shanghai's oldest garden. With the temple garden having been built no later than 1478 and having always remained an integral part of subsequent developments in what eventually became Qiuxiapu, it can at least claim to be the garden with the longest continuous history of the five located in Shanghai.

Each of the families associated with the early gardens that eventually merged to form Qiuxiapu bring their own contributions to the history of the site. The Gong Garden was established in 1502 by Gong Hong (龔弘), a member of the literati who is commemorated in the Hall of Virtuous Men in Suzhou's Surging Waves Pavilion. It is unclear whether Gong bought an existing garden or constructed a new one on the property, but the first description of the grounds was written by his associate. The records indicate that he lived in the compound for thirteen years after his retirement from officialdom. It is not certain whether other sites within the grounds, including the small garden behind the Gong house and the Sanyin Tang (三隱堂), were bought at the same time or later than Gong's first purchase of land here. If it was all purchased at the same time as Sanyin Tang, that would mark 1502 as the date of the Gong Garden's establishment, which is considered the latest possible date for its construction.

Gong reputedly took great pains to look out for the interests of the common people, promoting relief from taxes and financial aid in the years when flooding hurt the local economy. In 1535, less than thirty years after the death of Gong Hong, the Gong family met with financial difficulties of its own, and so sold the garden to the Wang (王) family, salt merchants from Anhui Province. Gong Xijue (龔錫爵), the great-great-grandson of Gong Hong, wanted to purchase the grounds back from Wang. The salt merchant told Gong Xijue that if Gong passed the imperial examination, the garden would be returned to him at no cost. When he passed the examination, the salt merchant kept his word, and the garden was again the possession of the Gong family. During the 1645 Three Massacres of Jiading, Gong refused to cooperate with Qing forces and was killed, so with no one in the Gong family to care for it, the family garden fell into disrepair. The salt merchant then returned to Jiading from Anhui to repair the grounds, naming the garden Qiuxiapu at that time. From that point on, there were no residential buildings included in the grounds, since the Wang family continued to live in Anhui. The garden remained a private leisure ground, but not a residential compound. The Wang family's fortunes later went into decline, and the garden was donated to the City God Temple in 1726.

The Shen Garden, on the east side of Qiuxiapu, was established in the early 1620s when Shen Hongzheng (沈弘正) developed a 10 *mu* (approximately 6.5 square metres) garden next to the Gong estate. The Shen Garden included the Fushu Tang (扶疏堂), the Liaoyan Tang (聊淹堂), a boat, an extensive rockery, and a winding calligraphy corridor among its features. Its main gate was once the place of honour for one of Dong Qichang's inscriptions. The Jin Garden, located in the northern part of the grounds, housed the Liuyun Ju (柳雲居), Zhi Fang (止舫), Jixia Ge (霽霞閣), and the Dongrong Guan (冬榮館). Both the Jin and Shen Gardens were absorbed into Qiuxiapu during the Qianlong period (1736–95). The Gong Garden was the oldest of the three and its history of passing hands the most colourful, and it is considered the main feature in the present-day grounds.

Jiading's City God Temple was first established in 1208–24 on a site near the present-day north gate of the garden. In 1307, the governor of Jiading ordered that the temple be moved to its present-day location. It was integrated into the complex of gardens early on, eventually absorbing parts (or all) of the family gardens into the temple complex for a period of time.

In modern history, Qiuxiapu has been a victim of several rounds of social upheaval. The garden was almost completely destroyed by Taiping forces in 1806 and seven years later, the people of Jiading undertook to rebuild the garden, opening up some commercial endeavours on the site that included a bazaar and, by 1888, a regular schedule of performances that were staged for the public. The Dai family, local Jiading residents, moved a private school into the back area of the garden, soliciting donations to restore the grounds.

In 1937, Japanese occupation forces took over Qiuxiapu, using it as their headquarters and a hospital. After the war, in 1945, a school again occupied the garden. The local government commissioned the school to care for the grounds, and it remained the site of Jiading's only public school until the 1960s.

In 1960, the townspeople renewed efforts to have the garden restored and protected, and a rebuilding effort initiated in 1963 forced the school from the grounds. Unfortunately, Qiuxiapu suffered another round of heavy damage during the Cultural Revolution. When that period ended, a new, costly restoration project was undertaken and completed in 1981. The restoration included the original gardens of the Gong, Shen, and Jin families, as well as the City God Temple. Historical records were used for the restoration works, with efforts being made to reproduce the original garden as nearly as possible.[2]

2. Ibid.

A Garden within a Garden and a View beyond a View

Having originally been four separate gardens now combined into one, Qiuxiapu embodies the notion of 'a garden within a garden and a view beyond a view' (園中之園，景外之景) that is so central to traditional Chinese landscaping. Qiuxiapu is generally thought of as having four main scenic areas based on the four original gardens. The Gong Garden is the present-day Taohua Tan area (桃花潭), the Shen Garden forms the present-day Ningxia Gallery area (凝霞閣), and the Jin Garden sits in the Qingjing Tang area (清鏡塘), while the City God Temple remains where it has been since 1307.[3] It is a labyrinthine garden with many hidden courtyards that can easily be missed in a single visit, so multiple tours of the grounds would be advisable for guests wanting to gain a full understanding of the garden's composition.

The layering of Qiuxiapu is achieved in two main ways. First, it is layered spatially, with one feature of the garden situated behind another, creating a system of screens that allows the garden to unfold slowly as one tours through it. Often while walking the grounds, the guest might be surprised by what lies on the other side of a doorway, whether it be a hidden courtyard, a rockery, a floral arrangement, or perhaps even the same courtyard in which the tour was first begun. In Qiuxiapu, perhaps more than in any other of Shanghai's gardens, it is not unusual to feel a bit lost with all the twists and turns in the paths that lead through the grounds.

A second form of layering is the equally winding system of intertextual references found in Qiuxiapu. The inscriptions on the doorways and the names of the features within the grounds are often taken from lines of old texts, selected by prominent figures

3. Ibid.

from later (but still ancient) history, and perhaps inscribed by a contemporary artist or scholar. In addition to these layers of literary references within the grounds, more layers unfold beyond the walls of the garden too, Qiuxiapu having frequently appeared in classical poetry since the Ming dynasty. The layers of meaning attached to the grounds mirror the spatial layering through which one physically experiences the garden, allowing body and mind to simultaneously make parallel journeys through the time and space occupied by the garden.

The waterways in Qiuxiapu are less expansive than those found in Qushui Yuan or Guyi Yuan, and less compact than the pond for which Zuibaichi is named. The garden exhibits more complex rockeries and more plant life than the other four gardens, and less water. Chen Congzhou explains that Taihu stones have been used in Qiuxiapu to create a feeling of water. Referring to Qiuxiapu as a 'dry garden', he says that in the back section of the grounds, the undulating rockery and sunken ground create an impression of a scene with a pool.[4] One of the main rockeries in the garden, the Guiyun Cave (歸雲洞), is inscribed with the words 'Dwelling of the Immortals' over its well-hidden north entrance. This inscription calls to mind the ancient beliefs upon which the first imperial gardens were built near Xi'an in the Qin dynasty, the belief that the immortals lived in the mountains. It was the desire to be near the immortals that drove early garden designers to begin experimenting with rock formations as a means of creating miniature mountains on one's own property, thus bringing the garden owner closer to the immortals.[5] The rocks were meant to look as if they had fallen straight from heaven, with an ideal rock being slim,

4. Chen, 68.
5. Peng, foreword.

translucent, and full of crevices and holes.[6] Furthering the celestial connection of this layered, twisting rock formation in Qiuxiapu, 'Guiyun Cave' sounds like 'Guanyin's Cave' in the local dialect.

Plants play a more prominent role in Qiuxiapu than in Shanghai's other gardens. The term *pu* (圃) in the garden's name refers to a cultivated field or orchard, as opposed to the more comprehensive *yuan* (園) in the names of Guyi, Qushui, and Yu Yuan, a character pointing to a residential garden. The character *pu* is used in 花圃 (*huapu*, 'flower bed'), 苗圃 (*miaopu*, 'nursery' or 'seedbed'), and 菜圃 (*caipu*, 'vegetable garden'). The Wang family, who chose the name Qiuxiapu, did not treat the garden as a residential compound, since they lived in their native Anhui. The choice of *pu* instead of *yuan* in the site's name seems to indicate its role as a leisure grounds or retreat rather than a residence. The notion of a *pu* as a pleasure grounds is reflected in the use of the character in 玄圃 (Xuanpu), the mythical fairyland on Kunlun Mountain, and in 瑶之圃 (Yaozhipu), the jade garden of the celestial ruler in paradise. The term *qiuxia* (秋霞) in the garden's name, meaning 'autumn mist', was chosen from a line in 'A Poetic Sequence on a Pavilion' (〈滕王閣序〉), a poem by Tang dynasty poet Wang Bo (王勃).

Literary Tradition

One of the main impulses behind the Jiangnan gardening tradition was the desire to create spaces for the cultured gentry to meet, compose new works of literature and art, and to put the results of their previous learning on display. In the history of every scholar's garden in the south, one will find mention of visits made to the site

6. Lou 2001, 168.

by esteemed writers, artists, and thinkers, some travelling from afar to participate in the appreciation and the continued development of a particular garden. The emperors Kangxi and Qianlong were frequent guests in the gardens of Suzhou, and their replication of southern-garden styles in the north is a subject often remarked on in studies of Chinese gardens. The fact that the plot of *The Dream of the Red Chamber* (《紅樓夢》) includes preparing a garden to receive visiting dignitaries has, likewise, been commented upon at length. The importance of the practice of leaving an inscription behind is also seen in both the novel and the imperial visits to Suzhou's gardens. In fact, it has been said that a garden is not really complete without such namings and inscriptions.[7] The visit of Liu Yunfang to Qushui Yuan and Dong Qichang to Guyi Yuan, mentioned in earlier chapters of this monograph, are other instances of this tradition of one scholar visiting another's garden.

Qiuxiapu's intricate connection with the literary arts made it a site that naturally drew much interest from touring members of the literati. An additional attraction of the grounds was the presence of a large contingent of the literati in Jiading during the Ming-Qing era. The 'four masters of Jiading'—Li Liufang (李流芳), Lou Jian (婁堅), Cheng Jiasui (程嘉燧), and Tang Shisheng (唐時升)—all frequented the grounds of Qiuxiapu during the Ming dynasty, using it as a site for writing, painting, and reciting poetry.[8] Several calligraphies and inscriptions in Qiuxiapu today commemorate such events from the garden's past. Hou Fang (侯訪), Ming dynasty author of the *Annals of Jiading* (《嘉定縣誌》), wrote a series of poems upon a visit that Xu Keqin (徐克勤), magistrate of Shaanxi

7. Lu, 067: 9.
8. 'Qiuxiapu', 《秋霞圃》，上海市地方誌辦公室，www.shtong.gov.cn/node2/node71994/node82435/node82438/index.html. Accessed on 29 October 2013.

Province, made to the grounds. This excerpt from that series paints a picture of the two scholars' meeting in the grounds:

和徐克勤先生移居金氏園詩五首

三十年前駐此曾，竹塘荷溆記分明。晴窗霞氣日常護，止舫煙波渾不驚。真隱寧須別大小，和風偏欲吹枯榮。閒來尹綠樓頭坐，柳葉梢邊看月生。

From *Five Poems in Response to Xu Keqin's 'Moving in to the Jin Garden'*

after three decades stationed far away
you still remember the bamboos on these banks
in the clear, sheltering breeze each day at twilight
the boat moored in the misty waters sits undisturbed
the hermit distinguishes between priorities
as the soft breeze stirs our longing
for the dried plants' next flourishing
and the next chance to sit idly in the pavilion
lazily ruling over the greenery from your upstairs room
while the willow branches watch the moon's rise

Years later, Zhao Yu (趙俞), the Qing dynasty magistrate in Shandong, visited the garden in 1688 and left this inscription for display in one of the garden's pavilions:

題柳雲居

林廬清灑見遺風，六世藏書在此中。聞説鄰家歌舞地，更無蝴蝶過牆東。

Inscription in the Willow and Cloud Residence

from the forest hut
through the mist spied
customs of old passed down

six generations
of collected books
lie herein
I have heard
the neighbours'
song and dance
but even the butterfly
rejects the invitation
lingering inside these walls

In his examination of the layers of meaning attached to Yu Yuan, classical gardening scholar Lu Shaoming describes the role such inscriptions play, saying, 'Although these various texts, images and materials often serve as artifacts in the garden, they show the feeling of peace and pleasure and link the objects and the courtyards into meaningful complexity [with] the scenery of the garden . . . [to be rehashed] by different users at different time[s].'[9]

In considering gardens, Western observers might be inclined to view them primarily as spatial entities. Traditionally, though, a Chinese garden was almost as much a text as it was a space. Lu writes, 'Various devices have been employed to depict Chinese gardens: maps, paintings, drawing[s], poem[s], novels, historical records and so on. Generally, Chinese scholars tend to use the language of the humanities (e.g., literature and art), rather than that of physical science, to elaborate on the complexity of Chinese gardens.'[10] Each garden rests on a foundation of art and literature as firmly as it does a foundation of earth and stone. In a garden such as Qiuxiapu, where its role as an intertextual space is foregrounded, one can virtually tour the grounds through literary references.

9. Lu, 067: 10.
10. Ibid., 067: 1.

A discussion of the grounds of Qiuxiapu generally begins with its central feature, the Taohua Pond scenic area, which includes more sights and hidden courtyards than any other part of the garden. Besides the pond itself, this area includes several prominent rockeries, such as the Guiyun Cave, the North and South Hills (北山，南山), the Hengqin Stone (橫琴石), the Sanxing Stone (三星石), and the Xianren Cave (仙人洞). The Ningxia Gallery, dating from the 1860s, is situated near the South Hill. The gallery originally stood in the Jin Garden before being absorbed into the Gong family's estate.

Two bridges skirt the southern portion of the pond, connecting the southern rockery to the buildings standing on the pond's eastern side. The Shequ Qiao (涉趣橋, 'Approaching Mirth Bridge') was named by Lou Jian in 1621, and it is one of the few Jiangnan-style bridges to bear this sort of inscription. The Three-Turn Bridge (三曲橋) is carved with the character 壽 (*shou,* 'longevity') and a bat pattern. The bat is often used in engravings in Jiangnan, usually depicted hanging upside-down. The term 蝠倒 (*fu dao,* 'upside-down bat') is a homophone for 福到, meaning 'good fortune has arrived'.

The Thatched Cottage (池上草堂) on the west side of the pond takes its name from the titles of two of Bai Juyi's poems: 'On the Pond' (〈池上篇〉, cited in Chapter 1) and 'Record of the Thatched Cottage' (〈草堂記〉). The hall resembles a boat, and therefore is also known as the 'Boat That Never Sails' (丹而不遊軒). A mirror inside the hall not only expands the space in the courtyard, but also highlights the contrasts between reality and its shadow or reflection,[11] while also emphasizing that the guest is a part of the scenery in the garden and is meant to participate in it. The hall

11. 'Qiuxiapu'.

contains several couplets, including this one by Zhou Chengzhi (周承志), penned in 1936:

> 池上草堂聯
>
> 軒窗傍水琴書靜
> 澗穀新晴草木香
>
> Thatched Hut on the Pond
>
> near the veranda's window, the melody of the water
> rising from the ravine, the fragrance of new growth

The nearby Osmanthus Veranda (叢桂軒) is a favourite viewing spot for many guests to Qiuxiapu. Each window in the pavilion offers a different view of the groves of osmanthus, maple, and bamboo planted among the rockeries surrounding the hall. The layout of the plants has been designed so that the scenery is vibrant in every season.[12] In early autumn, the fragrance of the osmanthus permeates the pavilion and the area surrounding it, its blossoms dropping like yellow snowflakes on the breeze later in the season. In imperial China, the phrase 'plucking the osmanthus flowers from the Moon Palace' (蟾宮折桂) was used in reference to passing the imperial examination. The plant's association with the moon comes from the fact that the object visible in the moon is thought to be an osmanthus bush. Legend surrounding the immortal Wu Gang (吳剛), a Sisyphus-like figure in Chinese mythology, tells of his banishment to the moon, where he continually chops the osmanthus tree, which possesses the power to immediately heal its own wounds.[13] The bamboo intermingled with the osmanthus in this section of Qiuxiapu maintains a vibrant greenery around the pavilion throughout the year, while also

12. Ibid.
13. Valder, 313.

creating a pleasing contrast between its straight, upright boughs and the curve of the windows of the pavilion.

On the north side of the pond, situated next to the Guiyun Cave, the Biguang Pavilion (碧光亭) faces water on three sides. With its excellent views of the fish and the lotus in the pond, it is a popular spot for celebrating the Mid-Autumn Festival. More of Zhou's poetry adorns the walls of the Biguang Pavilion:

> 碧光亭聯
>
> 快哉此亭翼然臨與池上
> 知者樂水從之宛在中央
>
> Pavilion in the Azure Glow
>
> how lovely the eaves of this pavilion
> and the pond in which they are mirrored
> the wise take delight in the water
> following its flow from here to the garden's heart

The Biwu Hall (碧梧軒), situated to the north of the Taohua Pond, is a four-sided hall that once served as the main reception room of the Gong family. The building has undergone many extensions and renovations over the years. The hall's name comes from a poem by Tang dynasty poet Du Fu (杜甫), 'parrots have left several grains of rice / a phoenix is perched on a branch of the green parasol tree' (香稻啄餘鸚鵡粒，碧梧棲老鳳凰枝). The deciduous parasol, or phoenix, tree is the only tree upon which the fabled phoenix is believed to alight, and so it was often planted near a scholar's study, as the bird was thought to bring inspiration.[14] Across the pond, the rockeries on the south bank are thought to resemble the twelve animals of the Chinese zodiac when viewed from the Biwu Hall.

14. Ibid., 287.

To the east of the Biwu Hall, the Guanshui Pavilion (觀水亭) includes calligraphy of the pavilion's old name, penned by Chen Qiucao (陳秋草). A couplet by Zhou Chengzhi, also in the Guanshui Pavilion, reads:

> 觀水亭聯
>
> 臨溪而漁稻熟魚肥信清美
> 憑闌遙矚天光雲影共徘徊
>
> Water-Gazing Pavilion
>
> overlooking the brook and ripe paddy fields
> our catch of fish plump and tasty
> leaning against a rail, we watch
> distant clouds linger in the lights

The Ningxia Gallery scenic area, occupying the eastern side of Qiuxiapu, also contains many important structures. Among the most notable are the Miju Corridor (覓句廊) and the Xiju Pavilion (洗句亭) in the northeastern side of the old Shen Garden. The corridor winds north to south, with the pavilion located at the northern end. Many engravings grace the walls, again bringing the literary connections in the garden to the foreground.

Another of Zhou Chengzhi's couplets is displayed in the Binzao Fengxiang Hall, which is located in the western part of the site of the former Shen Garden. Zhou's couplet reads:

> 賓藻風香室聯
>
> 芳草有情夕陽無語
> 流水今日明月前身
>
> Hall for Welcoming Guests
>
> in the floral fragrance
> the fervour of the setting sun
> wordlessly at rest

today's waters flow
predecessor
to the glory of tomorrow's moon

The Binzao Fengxiang Hall actually consists of two buildings combined into one. Standing across the pond from the Osmanthus Veranda, the two pavilions create a pleasing contrast, and their reflections in the pond add a sense of depth to the courtyard.

To the southeast of the Binzao Fengxiang Hall is the Xianyan Gallery (閑研齋). The Xianyan Gallery is the main attraction in the Ningxia Gallery scenic area. It is situated so as to allow for excellent views of the Taohua Pond, and is one of the oldest buildings in Qiuxiapu. Zhou's couplet describes the hall in this way:

閑研齋聯

窗含遠樹通書幌
風颭殘花落硯池

Inkstone Study

the sight of distant trees through the curtain
my scroll
the lingering aroma of fallen flowers on the breeze
my inkstone

The City God Temple sits in the southwest part of Qiuxiapu, with a third scenic area centring around it. An ancient well also sits between the main buildings of the temple complex. The City God Temple remains an active site of worship today, honouring the city god of Jiading, an early Qing dynasty scholar Lu Longqi (陸龍其). The official explanation for why he was made the city god of Jiading is that he was greatly admired by the townspeople and so, after his death, they invited his spirit back to dwell in the temple and act as the city's protector. An alternative version of the tale indicates that it was politically motivated, with the Qing emperor installing Lu as the local god so that the emperor could endorse his own ideas with the name of a respected local figure. Either explanation is plausible, and there is probably some truth in both.

The Qingjing Tang scenic area is built around the Qingjing Embankment. The waterways are the main feature of this section of the garden, with winding canals stretching out from a central pond. The banks of these waterways are lined with magnolias, maples, and pines, creating an ever-changing palette in the scenery surrounding the water. The greenery that surrounds the embankment in the spring is almost overwhelming visually. In the autumn, the same walk will lead through a striking blaze of red maples, making the site look surprisingly different in the two seasons. The maples found in this area of the garden, a relative of the hickory, are often used in Jiangnan landscaping to line streets or canal banks. Its Chinese name, 楓楊 (*fengyang*, 'maple poplar'), comes from the fact that the sound of the wind in its leaves resembles that of the breeze blowing through poplars. It is also known as the Money Maple (元寶楓, *yuanbaofeng*), signifying that its leaves are shaped like the gold ingots used in ancient China.[15]

The Sanyin Tang sits on the northwestern edge of the pond in the Qingjing Tang area. A local scholar, Wang Mingshao (王鳴韶), younger brother of the Jiading-born writer Wang Mingsheng (王鳴盛), penned this verse about the hall, capturing the sense of movement connected to the seasonal changes in this part of the garden:

三隱堂詩

堂名三隱傍江湄，鬥鴨欄邊樹影移。共說書櫥真有腳，寒燈夜誦小窗時。

15. Ibid., 322–23.

Refuge for Three Scholars

refuge for three scholars
in this riverside hall
the ducks' fighting is altered
by the shadows of its trees
the bookshelves sway
rearranging themselves in the room
and light trembles in the night
as we recite these tales by the window

The Sanyin Tang is one of the oldest buildings in the garden, and is known as one of the most notable scenes in all of Jiading.

The other structures in this portion of the garden include the Liuyun Residence (柳雲居), the Qiushui Hall (秋水軒), and the Qingsong Pavilion (青松嶺). The Qingjing area of the garden houses some of the most impressive varieties of trees collected in any of Shanghai's gardens, with bamboos, pines, and plum trees forming the core around which the rest of the plants have been arranged. The following verse by Dai Sigong (戴思恭), president of the Jiading Educational Association during the Republican years, captures a measure of the beauty of the plant life in Qiuxiapu:

清夏竹枝詞二首

其一：小山北聳樹陰籠，叢桂軒開面面風。
　　　薄暮一尊同挹爽，解衣磅礴興無窮。

其二：草堂池上坐船頭，高柳絲絲拂地垂。
　　　侵曉一壺佳茗瀹，石闌閑聽轉鶯兜。

Two Bamboo Poems on Summer's Purity

–1–
the lofty trees
on the hill's north side
enveloped in shadow
the hall opens to the breeze

rustling the osmanthus grove
dusk culminating
draws out new brightness
unveiling the feeling
of infinite repose

–2–

sitting on the boat
near the lakeside thatched hut
stroked by the willows'
long silky tendrils
at dawn, a singular beauty–
a pot of tea put to boil
while the rocks listen idly
to the orioles' anthem

6
Yu Yuan: Staging a Family Drama

Built when the development of Jiangnan gardens was at its peak, Yu Yuan (豫園) is not only the pride of Shanghai, but also one of the finest Ming dynasty gardens remaining anywhere in China. Construction of the garden began in 1559, making it a near contemporary of Suzhou's Humble Administrator's, Lingering, and Master-of-the-Nets Gardens, Wuxi's Jichang Yuan, and Hangzhou's Liu Zhuang and Jiang Zhuang. This explosion of garden development in the region reflected a social change, the rise of a merchant class who possessed the financial means to imitate the gardens that had been restricted to the aristocracy and literati prior to this time. But Yu Yuan's builder, Pan Yunduan (潘允端), had the classical training of a scholar. In fact, his family was known in their time for having two generations of scholars (his father and three uncles, along with Pan Yunduan and his brother) living in one household.[1] Pan's work in and writings about the garden reflect this background.

1. Tang, 11.

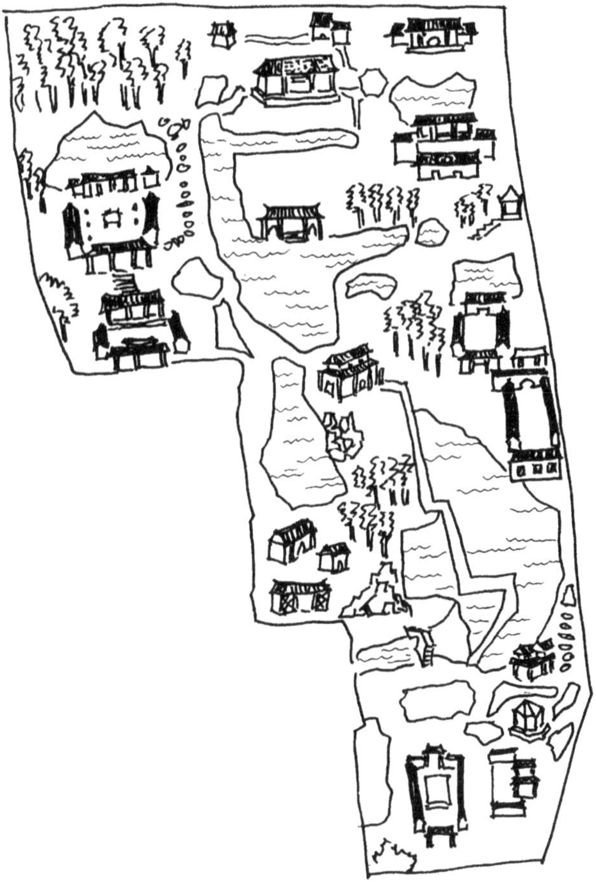

Figure 6.1
Layout of Yu Yuan (© 2015 Song Jie)

The Pan Family Garden

Pan Yunduan, having failed the imperial examination in 1559, determined that he would build a garden for his parents, naming it Yu Yuan, *yu* (豫) meaning 'pleasing and satisfying to one's parents'. Pan's father, Pan En (潘恩), was a famous man of letters in Shanghai, and many of his records of Ming dynasty affairs, including numerous town records, are still extant. The Pan family had long been prominent in the area and well educated, though they rarely occupied positions of real power. Pan En, however, served as an official at a fairly high post in local politics and was both influential and well-liked by the people of Shanghai. When the city wall was built by volunteers in 1553 to keep Japanese *wokou* (pirate) raiders out, Pan En not only assisted in the construction efforts, but also wrote an essay about the building process.

Six years after construction of the city wall was undertaken, Pan Yunduan began the laying of Yu Yuan. Three years after he had set out on this project, he again sat for the imperial examination, and this time he passed. The result was that he was assigned official posts for more than a decade, carrying out most of his civic duties as an official in Sichuan. Upon retiring, he returned to Shanghai in 1577, and after twenty years away, resumed construction of the garden. Author and China historian Florence Ayscough's translation of an excerpt from Pan's garden chronicle captures the spirit with which he threw himself into the job at this time: 'I thought only of the garden. I increased the size of the grounds, adding fifteen plots of land. I made seventeen pools. Furthermore, I bought many fields and devoted the entire revenue from these to beautifying the garden.'[2] With the increased experience he had gained during his

2. Pan, quoted in Ayscough, 222.

time away from Shanghai, Pan approached the development of the garden more systematically, claiming to apply 'Tao Yuanming thought' to his work. He hired Zhang Nanyang (張南陽) to design it, as Zhang was a leading expert in the laying of rockeries at the time. It took Pan more than ten years to complete the project, after which he personally penned the garden's chronicle, *Yu Yuan Ji* (《豫園記》). The record sounds somewhat like a tourist brochure, telling guests how best to view the site, since it was so expansive. He explains the history and the layout of the garden and points out the most scenic sites. His motivation for building the grounds is emphasized, saying that it was a form of filial expression meant to amuse his parents.[3] In the record, he also writes that the outcome was worth the wait and that the starts and stops were not problematic, as it allowed him time to think and also to accumulate sufficient financial resources for the project. In fact, he says that though he spent most of the family's fortune on the construction of the garden, he did so without regret. In 1590, he finally achieved all of his goals in the grounds, but his father had passed away in 1582, having not seen the garden in its finished state.

In its original form, Yu Yuan had many manufactured waterways, paths, rockeries, plants, and buildings. The grounds were designed to be a complex of different gardens with one centre, Leshou Tang (樂壽堂). The families of Pan Yunduan and Zuibaichi owner Dong Qichang had many close connections, often visiting each other's gardens. Dong wrote 'On Leshou Tang, for Pan Taihong's Birthday' (〈樂壽堂為潘泰鴻壽〉) about Yu Yuan's famous central hall. He describes the garden as being inside the city walls, home to old trees, a winding river, and mountains so well-made that they seemed natural. The mountains were piled

3. Tang, 13.

high, the waters clean and full of fish, and tall osmanthus trees stood on the banks of the ponds. The grounds were very large, with bridges crossing the canals and a waterfall flowing in one part of the garden. There was such a great variety of plants that Dong declined the impulse to start naming them, knowing there were too many to list.

When construction of Yu Yuan was completed, Wang Zhideng (王稚登), a renowned Ming dynasty calligrapher, inscribed the garden's name to be hung over the gate. To initiate the grounds, the Pan family invited the people of Shanghai to visit the garden before the family moved in. They hired members of the extended family to prepare meals, and for ten days there were various performances inside the garden staged for the public's entertainment. This event marked a long relationship Yu Yuan would have with the people of Shanghai.

The Uses of Yu Yuan

Florence Ayscough records something of an urban legend surrounding Yu Yuan in the 1920s. She writes, 'A persistent local tradition relates that the fame of Pan's garden reached the ears of the Emperor, who strongly disapproved that a subject should indulge in such magnificence, and that Pan found it advisable to make the whole place over to the public, who used it as the Temple of City Walls and City Moats; but the *Record* does not hint at such a thing.'[4]

What the record does say is that Pan Yunduan lived in Yu Yuan until his death. He had used much of his father's wealth to construct the garden, and even borrowed sums from other sources. His son continued this trend, squandering all the family's money

4. Ayscough, 226.

and eventually having no choice but to sell off parts of the family property. Other parts of the garden were donated to the Catholic Church, called in the record *Jingyi Tang* (敬一堂, 'Hall for Worship of the One'). It was Pan Yunduan's granddaughter who finally sold the last parcel of family property into the hands of outsiders, with Zheng Zhaolin becoming the owner of Yu Yuan in the early Qing dynasty.

Pan Fu (潘復), a descendant of Pan Yunduan, continued to live on a small patch of Pan land at the end of the Ming dynasty, and from there he helped defend the city wall when Manchu forces invaded. Upon the fall of the old city, Pan Fu led the family out of Shanghai, and most of the land that remained in the Pan name at this time was turned into a temple compound, with a Buddha being placed inside the temple in hopes that it would save the city from being ransacked. The Leshou Tang was destroyed, and all but a few small buildings and rockeries were left in a state of serious disarray. The local people used the former site of the glorious garden as a vegetable patch during the change of dynasties.

The City God Temple, though generally associated with Yu Yuan in the minds of contemporary guests to the city, did not originally stand on the site where it resides today. Instead, since the Three Kingdoms period (AD 220–80), it had faced the sea, its role being to provide protection from invasion by sea. The first city god was known as Huhai Shen (護海神, Sea-Protecting God) in records dating from 1292. The temple was moved to its present location in 1403.

Shanghai native Qin Yubo (秦裕伯) was invited to be the city god in 1344, after having served as an official in Fujian for twenty years. He had returned to Shanghai just at the beginning of the Ming dynasty, and when the new emperor asked Qing to take an official position in the new government, he refused, claiming that his mother was ill. When the emperor added the pressure of threats, Qing accepted the post and was sent to Nanjing. Though Qing never felt comfortable in the service of the new government, the emperor appreciated his efforts and, upon his death, built a shrine to him in Shanghai and named him the city god saying that, though he refused the role of the city's caretaker in life, it would be his in death. The Chenghuang Miao in Shanghai is the only City God Temple that honours two protectors in this way.

In the early days, the temple and Yu Yuan remained separate entities. The temple did not have a garden of its own when it was built, but in 1709, some funds were collected so that land could be acquired for a temple garden, and parts of Yu Yuan were purchased to reconstruct a garden on the dilapidated site, calling it Dong Yuan (東園, 'East Garden') or, alternatively, Nei Yuan (內園, 'Inner Garden'). The grounds were used mostly for ceremonies and performances, no longer serving as a residential compound, but as a temple garden. Not wanting to offend the emperor of the new dynasty, the garden was kept simpler, without any attempt being made to recreate the glory of the old Yu Yuan.[5] During the years of Kangxi's rule (1661–1722), perhaps encouraged by the emperor's love of Jiangnan gardens, a movement was initiated by the people of Shanghai that aimed to restore the old garden, but the effort failed due to a lack of funding. In 1760, the money for the project was finally raised and the land was purchased. Though the people wanted to rebuild the garden in its original form, no one was qualified to design it or oversee the work. In the end, they donated the garden to the City God Temple, naming it Xi Yuan (西園, 'West Garden'), giving the temple two gods and two gardens. It is these two gardens that form the modern-day Yu Yuan.

5. Tang, 13–14.

A poem by late Qing dynasty artist Wu Jiayou (吳嘉猷) describes the temple and the garden:

> 邑廟內園圖
>
> 欲問前朝事，殘碑不可尋。滄桑歷今古，城市有
> 山林。小閣春多雨，飛廊晝亦陰。笑他車馬客，
> 未解滌塵襟。
>
> A Picture of the Inner Garden in the City God Temple
>
> we seek knowledge of previous dynasties
> but the damaged stele cannot be found
> so many great changes in the world
> from ancient times until now—
> mountains and forests growing up
> even inside the city walls
> in the small gallery
> the spring brings many rains
> and daylight chases the shadows
> from the winding corridors
> we laugh at the gentry
> who arrive in fine carriages
> yet see no need
> to wash the dust from their garments

Though the temple expanded the garden's role as a public gathering space, the opening of the grounds to the public was not new. The Pan family, having initiated their residential compound by opening it for a public performance, had continued this tradition throughout their ownership of the property. In fact, of Pan Yunduan's journal entries from 1586 through 1601, nearly three quarters relate to various public performances that were held in the garden, with the Leshou Tang being the main venue for these events.

The sort of productions most often held in Yu Yuan were *kunqu* operas (Suzhou opera). Following a practice of cultured families from Suzhou, the Pans bought many children around the ages of seven or eight, trained them to sing or entertain, and made them a part of the *jiayue* (家樂, 'family musician') troupe. Pan Yunduan's journal also indicates that he purchased many trained *kunqu* performers and brought them to Yu Yuan to live with the rest of the family's musical team.

From just the first sixteen years after the garden's opening, there are over four hundred bills for musical performances of some kind. All weddings, funerals, and other family events included musical performances, adding to the public openings of the grounds. Classics such as the *kunqu* composition *Pipa Ji* (《琵琶記》) were staged in Yu Yuan. *Pipa Ji* took thirteen years to stage, with the twenty episodes from the piece being performed at different events. Pan Yunduan served as director for the performances, and he was well-known for breaking from convention and encouraging women to act in his productions.[6] It is apparent from Pan's own records that the garden served as a hub for much social activity in the city during this time, and that those events influenced the texture of Shanghai society.

Excerpts from a series of couplets displayed at the Guantao Lou, the first by Jiading native Yao Wennan (姚文枏) and the second by Qing dynasty writer Shi Fan (石範), capture the role Yu Yuan played in hosting many people during its peak:

觀濤樓聯

其一：且欣咫只窺岩壑
　　　便抱清暉就白雲

6. 'Yu Yuan', 《豫園》，上海市地方誌辦公室，www.shtong.gov.cn/node2/node71994/node82435/node82437/index.html. Accessed on 29 October 2013.

其四：得好友來如對月
　　　有奇書讀勝看花

From Tower for Watching the Billows

–1–
happy to spy even a small rocky mount
where the clear light soon brings white clouds

–4–
welcoming visiting friends
　as if greeting the moon
camaraderie
　shared over captivating books
exceeds even the beauty of the flowers

Many contemporary scholars say that walking through Yu Yuan is like walking through history.[7] The sights we encounter in the garden today call to mind the crowds of people—both members of the cultured gentry and the common people of Shanghai—who have gathered in this site for over four centuries.

Contemporary Yu Yuan

Yu Yuan today is not the same garden that Pan Yunduan built. It is much more compact than Pan's garden. In fact, we are not entirely sure of the exact boundaries of the Pan grounds, though we know they were expansive. What we know as Yu Yuan today is greatly influenced by the time the site spent serving as the temple garden, particularly in the separation of spaces such as the East and Inner Gardens. Lu Shaoming counts seven sub-gardens in the grounds of Yu Yuan, each enclosed by walls and with gateways connecting them.[8]

7. Tang, 11.
8. Lu, 067: 2.

The Leshou Tang, having been destroyed long ago, no longer rests at the centre of Yu Yuan. Taking its place of prominence is the Sansui Tang (三穗堂), a hall rebuilt in 1760. Its name, taken from a late Han dynasty story 'Grain Hanging from the Roofbeams' (〈梁上三穗〉), brings to mind thoughts of good fortune and abundance for those familiar with the story.[9] These Qing dynasty couplets hang in the hall:

三穗堂聯

其一：山墅深藏峰高樹古
　　　湖亭遙對橋曲波皺

其二：此即濠間非我非魚皆樂境
　　　恰來海上在山在水有遺音

Three Grains Pavilion

–1–
a mountain villa
amidst towering peaks
and ancient trees
a lakeside pavilion
faces a distant bridge
winding across the waves

–2–
here at this moat
the fish and I
all happily situated
on the sea[10]
in the hills and the water
our voices linger

In an extensive analysis of the site, Lu demonstrates the way that each courtyard in Yu Yuan offers a sort of subplot to the

9. Tang, 16.
10. Referring to Shanghai's name, which means 'on the sea'.

framing garden's main theme,[11] with the courtyard around the Sansui Pavilion centring around harvest and longevity.[12] The carved windows in and around the Sansui Tang are some of the most intricate in any of Shanghai's gardens, all carved with depictions of various grains. This is tied to the garden's overarching theme of respect for one's elderly parents in that it expresses wishes for their long life and renewed health. The Sansui Tang was originally used as a gathering space for reciting poetry, and Pan's original garden chronicle is displayed in the hall in a calligraphy penned by Pan Boying (潘伯鷹).[13]

Behind the Sansui Pavilion is the Yangshan Hall (仰山堂), first built in 1866. The second floor of the pavilion, called the Juanyu Lou (卷雨樓), is commemorated in this couplet:

卷雨樓聯

其二：樓高但任雲飛去
　　　池小能將月送來

The Rolling Rain Tower

–2–
not blocking the clouds' flight
this tall tower
this small pool
prodding the moon toward us

The Yangshan Hall faces one of the most famous scenes in the garden, an extensive formation of huang stones, topped by the Wangjiang Pavilion (望江亭), which looks out over the garden walls to the Huangpu River. The original rockery of huang stones excavated from Wukang in Zhejiang Province was laid by

11. Lu, 067: 7.
12. Ibid., 067: 10–11.
13. Tang, 16–17.

Zhang Nanyang. The rockery is 14 metres high, 60 metres wide, and 40 metres deep, with a zigzag flight of stairs for climbing the mountain hidden among the stones. Though the path is closed to visitors today, the rockery remains a stunning example of how stones are used to create miniature mountain ranges that look very like natural mountains.

Behind the rockery is the Cuixiu Hall (萃秀堂). With the hall situated at the foot of the rockery and against the garden wall, Chen Congzhou notes that when inside, one does not feel oneself in downtown Shanghai. 'With only a wall in between,' he writes, 'it seems as if it were a division between heaven and earth—an excellent example of "separating" which helps to bring out the effect in scenery construction.'[14] A Qing dynasty couplet describing the atmosphere at the Cuixiu Tang reads:

> 萃秀堂聯
>
> 花香入座春風靄
> 曙色凝堂淑氣濃
>
> Hall for Gathering Elegance
>
> a floral fragrance
> entering the spring mists
> dawn's light gathers in the hall
> a refined air

A double corridor leads from the rockery and Cuixiu Tang to the Wanhua Lou (萬花樓). A pair of Qing dynasty couplets is displayed at the Wanhua Lou, the first by Xu Yongzhang (徐永章) and the second by Zheng Banqiao (鄭板橋):

14. Chen, 167.

萬花樓聯

其一：桂馥藍芬水流山靜
花明柳媚日朗風清

其二：春風放膽來梳柳
夜雨瞞人去潤花

Floral Tower

–1–
fragrance of osmanthus flowers
beside the blue waters
flowing through the silent hills
bright flowers and willows
wave in the charming light
and the clear pure breeze

–2–
spring winds
boldly comb willow strands
while night rains
hide the sleek flowers

Lu's analysis of the Wanhua Lou courtyard is helpful for gaining a deeper appreciation for the sub-theme expressed in this section of Yu Yuan. The overt message contained in the building's name and the motifs depicted there point to the abundance of flora in the grounds, particularly the display of trees and flowers arranged as a centrepiece on the opposite bank of the pool situated in this courtyard. A more subtle expression of an embedded theme comes to light when one learns that the gingko tree in front of the Wanhua Lou was planted by Pan En, Pan Yunduan's father.[15] The gingko, the sole survivor of a once-abundant genus native to the Jiangnan region, is a leafy tree that will turn bright yellow every autumn, even in locales where the temperature never drops to

15. Lu, 067: 8.

freezing, making it an ideal ornamental plant for emphasizing the change of seasons in the grounds of a garden. In texts dating from as early as the Han dynasty (202 BC–AD 220), the gingko has been revered as a holy tree that was often planted in temple grounds, and by the eleventh century, it featured widely in Chinese poetry and paintings. The trees were revered mainly for their longevity, symbolizing hardiness and resilience.[16]

The path across the Wanhua Lou courtyard leads to Yu Yuan's renowned dragon wall. The garden contains nine dragons in total[17] with the dragonhead atop this portion of the wall being the most distinct and most famous. The dragons, symbolizing immortality, serve as another thematic thread woven into the garden.

The area on the other side of the gate resting beneath the dragonhead contains the Dianchun Hall (點春堂). In it are many relics from the Small Sword Uprising of 1853, which used Yu Yuan as a sort of headquarters for their activity. Many of the buildings around the Dianchun Hall were occupied by the Small Swords Rebellion, being used mainly as offices.[18] The stage facing the hall is a reminder of the earlier role that the garden played as a venue for performances in Pan's lifetime.

The area next to the Dianchun Hall courtyard is yet another sub-garden within Yu Yuan. Passing through a small bamboo grove, the path into this courtyard leads to the Jiushi Hall (九獅軒), situated on the north end of the garden's largest pond. From the porch of the study, the view of the courtyard shows a second larger pavilion on the left and a small gazebo in the distance on the right bank. At the opposite end of the pond is a moongate. When

16. Valder, 72–74.
17. Lu, 067: 10.
18. Tang, 19–20.

the breeze stirs, the willow branches stroke the edges of the pond, echoing the ripples of the carp swimming beneath its surface.

Walking along the path towards the smaller gazebo leads the guest around the pond and to a rockery built into the wall. A walk through the narrow cave in the rockery reveals the surprising fact that the circuitous path into the garden has brought the guest back to the starting point, for the main entrance lies on the other side of the wall and rock formation. On the opposite bank is the Huijing Lou (會景樓), a two-storey structure rebuilt in 1870. It once offered views of the whole garden, though the hall is not open to guests today.

Next to the moongate at the end of the pond is the Yuhua Hall (玉華堂). A replica of the building that once served as Pan Yunduan's study, the Yuhua Hall now exhibits Ming dynasty furnishings. The southern opening of this four-sided pavilion looks out onto the Yu Linglong (玉玲瓏), a Taihu stone relic from the Song dynasty. Its slimness and the water tracks covering this stone from top to bottom make it unique. It is said to be over a thousand years old. This stone, along with Suzhou's Duanyun Feng (端雲峰) and Hangzhou's Zhouyun Feng (縐雲峰), is one of the three primary examples of a Taihu stone in the region.[19] The stone already had a long history as a garden piece when it was given to Pan Yunduan, who moved it from Pudong's Nan Yuan to Yu Yuan.[20]

Beyond the Yuling Long, another courtyard leads to the Inner Garden. The Inner Garden contains many features that seem to be miniatures of those housed in the grounds framing it. One of the most intriguing is a dragon atop the wall, which appears to be poised to swallow a three-legged toad. Florence Ayscough reports that there were many theories in her day concerning this feature

19. Ibid., 21.
20. Ibid., 22.

in the garden. Her own belief was that the toad, a symbol of the earth, stood as a tiger in the traditional tiger-dragon motif, with the dragon standing for heaven. Ayscough reads this as a depiction of struggle or conflict. If, she says, the dragon swallows the toad, it must disgorge its meal after consuming it, or if it grasps the toad, the toad will swell in its grip until it lets go. The fight, then, is unending, pointing to the notion that neither heaven nor earth can defeat the other.[21]

Inside the Inner Garden is a reconstruction of the Pan Mansion with an ancient stage that was built in 1888 and is known as the premier stage in Jiangnan gardens. Its facade is decorated with wood carvings and gold paintings, and its circular ceiling has a mirror at its centre to enhance the acoustic effects. On the second floor of the building on the opposite end of the courtyard from the stage is a box reserved for VIPs in the audience, a wide room with tables where tea would have been served during a performance. The courtyard offers some idea of the opulence and pageantry that must have accompanied the performances hosted by the Pan family.

Chen Congzhou notes that Pan's mansion gets less attention today than the garden or bazaar at Yu Yuan. It was, however, well-known during the Ming dynasty. Situated a short distance from the garden, it was the largest mansion in Shanghai at the time. Chen cites records of the grandeur of the Pan mansion, stating that it offers 'collateral evidence of the dimensions of Yu Yuan in those days'.[22]

For many guests to Shanghai, the garden at Yu Yuan can almost be swallowed up by the retail bazaar that surrounds it. It is not uncommon to hear negative comments about the commercialism of the area, often suggesting that the shops have

21. Ayscough, 234.
22. Chen, 113.

somehow diminished the sense of tradition attached to the site. Such thinking, however, overlooks the nature of the garden as a venue often opened for public gatherings in Pan's day, and later as a public gathering site when it served as the temple compound.

Not only can the bazaar at Yu Yuan be seen as a part of this tradition, but the commercial aspect is also not nearly as new a phenomenon as many might imagine. Almost immediately after Yu Yuan became the temple garden, shops, banks, and restaurants were opened within the grounds. Funds generated from these enterprises were used to rebuild the garden. It became not only a public gathering place, but also a site that was publicly owned and maintained. Since at least 1860, the temple, garden, and bazaar have been lumped together as a complex hosting amusement and entertainment for the public, building on Pan's love of hosting public events when he owned the garden.

While it is easy to be critical of commercialization and rampant consumerism in ancient sites, in Yu Yuan, the bazaar is not a departure from tradition. For some, the lights may be too bright, the crowds too large, the noise too great, and the prices too high, but the truth is that Yu Yuan has managed to maintain a delicate balance between scholarly elegance and gaudy consumerism from its earliest days. It has always embodied the *re'nao* (熱鬧, 'lively, vibrant, and noisy') atmosphere so well-loved in public gathering places in China, even as it participated in the rich cultural heritage of the Jiangnan garden. From its inception Yu Yuan was built to host dramas, and no drama has been more emphatically played out in Yu Yuan than those of real human lives. It has served as the stage on which the drama of the Pan family and the drama of the people of Shanghai have been performed.

7
Conclusion

It is commonly held today that Chinese gardening is an extinct art form. The logic behind this thought is based on two facts. First, an art form that requires three centuries to come to fruition is simply not possible where property ownership is only on a lease basis, with leases being measured in decades rather than centuries. However, this line of thinking overlooks the point that many classical gardens—including four of the five that remain in Shanghai—were publicly-owned sites. Each of the four in Shanghai functioned for a time as its respective City God Temple grounds, and Qushui Yuan was never privately owned, having been a part of the temple compound from the beginning. This fact alone challenges the assumption that private ownership is a necessary prerequisite for the development of a classical Jiangnan-style garden. History has already proven that public funding and maintenance can initiate and sustain the development of a public garden.

A second reason many believe that the construction of Chinese gardens is a thing of the past is the fact that there is no longer a literati class, and so the very thought system that traditional gardening was based upon has passed completely from existence. It is important to remember, as Cai Yanxin points out, that 'to many scholars and literati, their material and spiritual lives were inseparable from the landscape gardens. In a landscaped garden, one could also feel the emotions and interests of these individuals.' This would seem to support the notion that Chinese gardening is no longer possible in a contemporary context. But on the other hand, Cai continues, 'A landscape garden is not merely composed with artificial hills, rivers, and pavilions. It is a work of art that is put together with different artistries that still exist in China's culture and society.'[1] The question lies, perhaps, in whether today's continuation of the art forms with which the Chinese garden has always been so closely associated is enough of a foundation for the continued building of scholars' gardens.

In Shanghai, only five classical gardens receive official recognition. It is generally understood that of the dozens of gardens in Shanghai during the Ming-Qing era, these are the only ones that have survived the passage of time. The choice to limit this study to these five gardens, then, has not been arbitrary, following the oft-mentioned fact in the literature of the Shanghai Municipality, both in English and Chinese, that these are the only five remaining classical gardens in the city—though this does overlook, at the least, Kezhi Yuan in Zhujiajiao, a water village in Qingpu District. In addition, there have been some efforts to rebuild other gardens, the most notable being the Tan Garden in Nanxiang, located a short distance from Guyi Yuan. The Tan Garden was built on the site where its Ming dynasty predecessor originally stood and has been rebuilt based on early records and employing the finest craftsmanship from Suzhou. When the project was initiated, there were no remains of the original garden, and so the beginning point for the work was simply the words of the garden chronicle. Chen

1. Cai, 127–28.

Congzhou suggests that a distinction be made between 'restoration' and 'rebuilding' of classical gardens, commenting that a full restoration requires decades of research into the literature and art associated with a particular garden, in at least equal measure to the attention given to the layout of the land and the remains found there.[2] The Tan Garden clearly presents itself as a rebuilding, an attempt to recapture—and perhaps even reinterpret—the ancient garden for a modern audience.

Such efforts suggest that there is a foundation for the continuation of China's gardening tradition, with some adaptations perhaps necessary to accommodate modern sensibilities. Further evidence that traditional gardening can thrive in a contemporary setting can be seen in the Japanese gardening tradition, which grew out of Chinese practices. In Japan, classical landscaping was not interrupted by social upheaval in the same way that it was in twentieth-century China, and the styles and forms of Japanese gardens have continually grown and developed over the years. Chen, who looked to Japan for some inspiration in his own restoration works, writes, 'Before the Meiji Reformation, the Japanese mostly learned from China; and after the Meiji Reformation, they modeled themselves on Europe, and later on the United States. But all the time, their architecture and garden designing have kept to the Japanese national style, or "the Japanese flavour," so to speak.'[3] Perhaps it is possible for 'the Chinese flavour' to likewise continue in a revived gardening tradition.

The question that naturally arises is what this 'Chinese flavour' might look like. How might it remain rooted in tradition, and yet take on new forms? Perhaps an answer lies in a renewed appreciation of the various roles the gardens played in the culture of their

2. Chen, 90.
3. Ibid., 169.

day and how that might be duplicated (and ultimately expanded) today. For example, the gardens have long served as sites for receiving visiting dignitaries, and also as repositories for commemorating those visits through the various inscriptions left by the guests to the grounds. In Yu Yuan, an inscription by Jiang Zemin, left to commemorate his 1999 visit to celebrate the garden's 400th anniversary, stands just inside the garden's entrance, near the Sansui Tang. The inscription reads 'Famous Garden on the Sea' (海上名園).[4] Such events follow in a long tradition, and their continued use for such gatherings opens the possibility for contemporary interactions and expressions that can help in an ongoing development of these cherished sites.

The main purpose of the classical gardens, however, was not to receive women and men with some political clout. The real VIPs in the days of the classical gardens were artists of all sorts, including painters, poets, calligraphers, carvers, and others of the literary or artistic professions. Garden culture was inseparable from the other arts, and was often used to host events centred around the arts. Cai Yanxin comments on this interaction between art forms, saying, 'Chinese classical arts, especially poetry and paintings, had greatly influenced the designs of landscaped gardens. On the other hand, poets and artists also obtained their creative inspirations from the tranquil sights offered by a beautifully landscaped garden. Arts and landscaped gardens complement each other in their glories.'[5] Classical gardens were the sites most frequently chosen for poetry contests, drama performances, and calligraphy and painting exhibitions, along with celebrations of festivals, chess matches, and lantern displays.[6]

4. Tang, 10.
5. Cai, 127.
6. Hardie, 7.

Perhaps the revival of gardening culture, then, ultimately lies in the hands of today's artists and poets. The clusters of artists with their sketchpads seen in the grounds of all gardens today offers hope that there is still the possibility for interaction between gardens and other forms of artistic expression. Along this same line, the display of contemporary art to be found in each of Shanghai's gardens is another encouraging sign that the administrators of the grounds are aware of the need for continued interaction between the garden and other art forms. In 2010, the grounds at Kezhi Yuan in Qingpu served as the set for a production of the classical *kunqu* opera *The Peony Pavilion*, reviving the ancient tradition of staging public performances in the grounds of a Jiangnan garden.[7] If efforts such as

Figure 7.1
Scene from Qushui Yuan (© 2015 Song Jie)

7. Ibid., 79–81.

these could continue to be extended beyond horticultural displays and exhibitions of paintings to include poetry readings or workshops, more music and theatre performances—or even offering spaces for working artists in the manner that Qushui Yuan housed Shen Shoudong's office in the early twentieth century—it would mark a renewal of the tradition of using the gardens as a centre for art and culture. Paintings and poetry taking the scenes of the garden as their subject would become more common and more work of this nature would appear in literary journals and anthologies, furthering the reach of gardening culture. This 2013 verse by the Japanese contemporary poet Miho Kinnas, commemorating her visit to Qushui Yuan, serves as an example of the sorts of responses that might be made to the gardens today:

> Qushui Yuan
>
> Orange Osmanthus petals
> fall on the rocks, roofs, and water
> The weak sun from a slit in the clouds
> reflects the pond dancing
>
> upside down on the eaves
> Stepping along and up
> the stairs we arrive at
> the poet's study
>
> The windows all around
> view the garden see the gazebo
> Two strangers today
> cramp the regulars
>
> Carps jump and the wind runs
> Listening and not listening
> Conversations in foreign tongues
> This afternoon passes[8]

8. Miho Kinnas, used with permission.

As the tradition of gardens interacting with the other arts continues to grow, it would be a natural extension for more grounds to be designed and opened to the public as centres for the arts, just as happened in earlier generations in China. New forms and styles would develop to express the modern 'Chinese flavour' similar to the 'Japanese flavour' found in contemporary gardens in Japan. And one need only look to I. M. Pei's design for the New Suzhou Museum, opened in 2006, for an idea of how classical gardening techniques can be integrated with contemporary architectural designs to create public spaces that are engaging on both the utilitarian and aesthetic levels.

The role that Chinese gardening might take in the future is one that I hope would be considered and explored not only by designers, but also by artists in other fields who might promote the continued interaction between garden and art. For now, though, the classical gardens of Shanghai remain as a remnant of a bygone era. It is my hope that this monograph will help other viewers of the gardens better understand and appreciate what has gone into their construction and the role they have played in the history and development of Shanghai's culture and, if the day comes, that these sites are more integrated into the contemporary arts scene, to be ready to engage in the tradition as part of the garden culture community.

It is my belief that forging a future direction for gardens in contemporary China depends on our ability to engage with their past. In fact, in doing so, we are already participating in the tradition of Chinese landscaping, which always looked to the past as it built for its own time and the future. Several mentions have been made in this volume of Tao Yuanming's essay 'The Land at the End of the Peach Grove'. This piece, written in the fifth century, never specifically mentions a garden, but it has long been held as the encapsulation of the thought on which later gardening culture rests. Perhaps, then, it is best to close this discussion of Shanghai's gardens not with a look forward, but with a look back to Tao.

The Land at the End of the Peach Grove

During the Taiyuan period of the Jin Dynasty, a fisherman from Wuling set out in his boat to ply his trade. He followed the turns of the stream, oblivious of the distance he had travelled, and suddenly found himself in the middle of a peach grove lining the banks for several hundred metres. He noticed a complete absence of other varieties of foliage—all around was the fragrance of wild flowers, and the ground was carpeted with petals from peach blossoms. In amazement, the fisherman continued his journey, hoping to discover the end of the grove, which turned out to lie at the head of a stream.

The stream ran alongside a mountain, in which was a crag lit by a shaft of light. The fisherman abandoned his skiff and entered the narrow opening, barely wide enough for a man to pass through. He proceeded another dozen paces or so and found that the tiny passage opened onto a breathtaking vista. He could hardly believe the sight that met his eyes—a plain dotted with orderly houses, fertile fields, and ponds lined by mulberry and bamboo trees; roads crisscrossing the land, filled with the sound of farm animals; and men and women working the fields wearing clothes identical to those worn by labourers on the other side of the cave. He noticed that old and young alike seemed to move about with an easy freedom.

Upon seeing the fisherman, the villagers were astonished. When they asked where he had come from and what he was there for, he answered all their questions. They invited him to stay a while, and promptly set about preparing wine and meat for a grand feast. As word of the stranger's arrival spread through the village, many came to see him with their own eyes. The men explained that their ancestors, during the turbulent times of the Qin dynasty, had taken their wives and children and fled from the political unrest to this idyllic spot, completely cutting off contact with the society outside of their quiet oasis. They asked how things had gone on since those times, and the fisherman found that they knew nothing even of the Han kings, much less the Wei and Jin rulers who came after. When he had told them all he knew, they were dismayed by the level of their own ignorance of the affairs of which he spoke. The fisherman enjoyed the hospitality of the people, staying several days with them. When he finally bade them farewell, the villagers said, 'There's no need to let anyone else know about us and our life here.'

Returning through the passage, he found his boat moored where he had left it and made his way back home, carefully marking the path as he went. Upon reaching the city, he reported to the prefect all that he had seen. The prefect immediately mustered his troops and sent them out with the fisherman, tracing his marks in hopes of reaching the land that lay beyond the peach grove. Search as they might, they could not find the way there and were soon lost. They gave up the quest and returned to the city.

Liu Ziji of Nanyang, a hermit with high moral values, heard the tale and set out in search of the Land at the End of the Peach Grove, but to no avail. He never found the path to that secret paradise and shortly after his quest, he died of illness, so no one dared to go searching for that land again.[9]

桃花源記

晉太元中，武陵人捕魚為業。緣溪行，忘路之遠近。忽逢桃花林，夾岸數百步，中無雜樹，芳草鮮美，落英繽紛。漁人甚異之。復前行，欲窮其林。

林盡水源，便得一山。山有小口，彷彿若有光，便舍船，從口入。初極狹，纔通人；復行數十步，豁然開朗。土地平曠，屋舍儼然，有良田美池桑竹之屬。阡陌交通，雞犬相聞。其中往來種作，男女衣著，悉如外人。黃髮垂髫，並怡然自樂。

見漁人，乃大驚，問所從來。具答之。便要還家，設酒殺雞作食。村中聞有此人，鹹來問訊。自雲先世避秦時亂，率妻子邑人來此絕境，不復出焉，遂與外人間隔。問今是何世，乃不知有漢，無論魏晉！此人一一為具言所聞，皆歎惋。餘人各復延至其家，皆出酒食。停數日，辭去。此中人語雲：「不足為外人道也。」

9. Author's translation, first published in the Shanghai Arts Community 2011 exhibition, *Things That Disappear*.

既出，得其船，便扶向路，處處誌之。及郡下，詣太守說如此。太守即遣人隨其往，尋向所誌，遂迷，不復得路。

南陽劉子驥，高尚士也，聞之，欣然規往。未果，尋病終。後遂無問津者。[10]

10. 陶淵明，《桃花源記》，http://www.lovegod123.com/love/love-48.htm. Accessed on 2 December 2013.

Glossary

Terms to Note

City God Temple (城隍廟): located in each town and city, a temple dedicated to a prominent local figure who, after death, was believed to serve as spirit protector of the city and its inhabitants

haipai (海派): 'local colour' or 'fusion'

huaniao hua (花鳥畫): Chinese bird-and-flower painting

huicui (薈萃): collage or bricolage

Jiangnan (江南): the region south of the Yangtze Delta

shanshui hua (山水畫): Chinese landscape painting

wenren (文人): literati class

Features in a Classical Garden

artificial mountains (假山): rocks piled to imitate mountains and mountain ranges in miniature form; usually built on the edge of a waterway

covered walkways (廊坊): roofed paths that wind through the garden, framing and screening the views as they twist through the grounds

dragon wall (龍牆): a garden wall whose top is patterned to look like a dragon's body, featuring a dragon's head (or several dragons' heads) at some point in the garden

four-sided pavilion (四面廳): a structure open on four sides, framing four different views of the garden for those situated inside the pavilion

galleries (軒): the word implies a spacious, lofty area; galleries are often situated in high, open spaces and incorporated into the scenery gate tower (門樓): traditionally, the gate tower was the only structure with a prescribed orientation, being aligned with the great hall and situated in relation to it

gazebo (榭): gazebos are made to 'borrow' or take advantage of the scenery; usually situated beside water or plant arrangements

great hall (廳堂): one of the largest buildings in a garden; in ancient times, it was preferable for the great hall to either be three or five spans, depending on the space allowed for the garden

huang stone (黃山石): large, flat stones stacked up to construct various parts of a Chinese garden, especially miniature mountain ranges

marble boat (石舫): pavilion built to look like a boat, with three sides over the water; the best create a feeling of movement

pavilion (亭榭): shady gazebo, usually among plants or flowers, or on the edge of a pond

penthouse (簷): built up against a cliff face, or buildings which are not complete in themselves but use a cliff face to form part of their structure

study (書房): studies can come in a variety of shapes; usually the inside and the outside of the structure are not strictly delineated

Taihu stone (太湖石): limestones taken from Lake Taihu and displayed in gardens, often as a single piece

terrace (台): the word means support; garden terraces may be made from rocks piled high with a flat surface on top or constructed of flat planks laid on a high wooden framework, with no buildings on top

tower (樓閣): the tower should be situated behind the great hall, amid hills and water

Bibliography

Primary Sources

Ayscough, Florence. *A Chinese Mirror.* Boston and New York: Houghton Mifflin Company, 1925.

Cai, Yanxin. *Chinese Architecture.* Translated by Andrea Lee, Selina Lim, and David Gu. Beijing: China Intercontinental Press, 2010.

Chen, Congzhou. *On Chinese Gardens.* Shanghai: Shanghai Press and Publishing Development Company, 2008.

'Dong Qichang (1555–1640) and Art-Historical Painting.' http://www2.liu.edu/cwis/cwp/but06/hillwood/chinese/essays/no_essay_04.html. Accessed on 29 October 2013.

Ehrlich, Gretel. *The Future of Ice: A Journey into Gold.* New York: Vintage Books, 2004.

Fung, Stanislaus. 'Longing and Belonging in Chinese Garden History.' *Perspectives on Garden Histories* (1999): 205–19. Print.

Hardie, Alison. *Chinese Garden Pleasures: An Appreciation.* Shanghai: Shanghai Press and Publishing Development Company, 2014.

Ji, Cheng. *The Craft of Gardens: The Classic Chinese Text on Garden Design.* Translated by Alison Hardie. Shanghai: Shanghai Press and Publishing Development Company, 2012.

Leigh, Brandi. 'Dong Qichang.' *The Art History Archive—Chinese Art,* 2008. http://www.arthistoryarchive.com/arthistory/asian/Dong-Qichang.html. Accessed on 29 October 2013.

Lou, Qingxi. *Chinese Gardens.* Translated by Zhang Lei and Yu Hong. Beijing: China Intercontinental Press, 2003.

Lu, Shaoming. 'From Syntax to Plot: The Spatial Language of a Chinese Garden.' University College London, Bartlett School of Architecture. http://www.sss7.org/Proceedings/03%20Spatial%20Analysis%20and%20Architectural%20Theory/067_Lu.pdf. Accessed on 6 November 2013.

'Qingpu Museum.' http://museum.shqp.gov.cn/gb/content/2009-02/23/content_237430.htm. Accessed on 3 November 2013.

'Six Dynasties Calligraphy.' http://depts.washington.edu/chinaciv/callig/7calsixd.htm. Accessed on 3 November 2013.

Valder, Peter. *The Garden Plants of China*. Portland, OR: Timber Press, 1999.

Yan, Junming. *Enchanting Nanxiang*. Shanghai: Shanghai Cultural Arts Publishing, 2010.

Ye, Shengtao. 'The Suzhou Garden.' Author's translation, original text available online at http://www.5156edu.com/page/07-06-07/24814.html.

Zheng, Yuanxun. 'A Personal Record of My Garden of Reflection.' Translated by Duncan Campbell. *Asian Studies Institute Translation Paper* (November 2004): 1–22. Print.

白居易。《池上篇》。http://www.haoshici.com/Baijuyi46335.html. Accessed on 2 November 2013.

《古猗園》。上海市地方誌辦公室。http://www.shtong.gov.cn/node2/node71994/node82435/node82439/index.html. Accessed on 29 October 2013.

《近代名園記述選錄》。上海市地方誌辦公室。http://www.shtong.gov.cn/node2/node2245/node69854/node69866/node69883/index.html. Accessed on 2 November 2013.

樓慶西。《中國古建築二十講》。北京：生活・讀・新知三聯書店，2001。

彭一剛。《中國古典園林分析》。北京：中國建築工業出版社，1986。

《秋霞圃》。上海市地方誌辦公室。http://www.shtong.gov.cn/node2/node71994/node82435/node82438/index.html. Accessed on 29 October 2013.

《曲水園》。上海市地方誌辦公室。http://www.shtong.gov.cn/node2/node71994/node82435/node82440/index.html. Accessed on 29 October 2013.

唐名聲。《海派園林》。上海：文匯出版社，2010。

陶淵明。《桃花源記》。http://www.lovegod123.com/love/love-48.htm. Accessed on 2 December 2013.

吳義。《秋霞圃志（重修本）》編纂小記。http://www.shtong.gov.cn/node2/node70393/node70403/node81552/node81554/userobject1ai105092.html. Accessed on 29 October 2013.

《豫園》。上海市地方誌辦公室。http://www.shtong.gov.cn/node2/node71994/node82435/node82437/index.html. Accessed on 29 October 2013.

《園林楹聯、詩詞選錄》。上海市地方誌辦公室。http://www.shtong.gov.cn/node2/node2245/node69854/node69866/node69882/index.html. Accessed on 3 November 2013.

《醉白池》。上海市地方誌辦公室。http://www.shtong.gov.cn/node2/node71994/node82435/node82441/index.html. Accessed on 29 October 2013.

Secondary Sources

Chambers, William. *A Dissertation of Oriental Gardening*. London: W. Griffin, 1772; rpt.: Farnborough: Gregg, 1972.

Chung, Wah Nan. *The Art of Chinese Gardens*. Hong Kong: Hong Kong University Press, 1982.

Fung, Stanislaus, and Mark Jackson. 'Four Key Terms in the History of Chinese Gardens.' In *Proceedings of the International Conference on Chinese Architectural History*. Hong Kong: Chinese University of Hong Kong, 1995.

Graham, Dorothy. *Chinese Gardens: Gardens of the Contemporary Scene; an Account of Their Design and Symbolism*. New York: Dodd, Mead and Co., 1938.

Han, Pao-te. *The Story of Chinese Landscape Design: External Forms and Internal Visions*. Translated by Carl Shen. Taipei: Youth Cultural Enterprise, 1992.

Howard, Edwin Laclede. *Chinese Garden Architecture: A Collection of Photographs of Minor Chinese Buildings*. New York: Macmillan, 1931.

Inn, Henry, and Shao-chung Lee, eds. *Chinese Houses and Gardens*. Honolulu: Fong Inn, 1940.

Johnston, R. Stewart. *Scholar Gardens of China: A Study and Analysis of the Spatial Design of the Chinese Private Garden*. Cambridge: Cambridge University Press, 1991.

Kerby, Kate. *An Old Chinese Garden: A Three-Fold Masterpiece of Poetry, Calligraphy and Painting, by Wen Chen Ming, Famous Landscape Artist of the Ming Dynasty*. Shanghai: Chung Hwa Book Co., 1922.

Keswick, Maggie. *The Chinese Garden: History, Art and Architecture*. New York: Rizzoli, 1978.

Lu, Andong. 'Lost in Translation: Modernist Interpretation of the Chinese Garden as Experiential Space and Its Assumptions.' *Journal of Architecture* 16 (4) (2011): 499–527.

Markbreiter, Stephen. 'Yu Yuan: A Shanghai Garden.' *Arts of Asia* 9 (6) (November–December 1981): 99–110.

McDowall, Stephen. 'In Lieu of Flowers: The Transformation of Space and Self in Yuan Mei's (1716–1798) Garden Records.' *New Zealand Journal of Asian Studies* 3 (2) (December 2001): 136–49.

Minford, John. 'The Chinese Garden: Death of a Symbol.' *Studies in the History of Gardens and Designed Landscapes* 18 (3) (Autumn 1998): 257–68.

Qiao, Yun, ed. *Classical Chinese Gardens*. Hong Kong and Beijing: Joint Publishing Company Ltd., 1982.

Siren, Osvald. *Gardens of China*. New York: Ronald Press, 1949.

Strassberg, Richard. 'Mirrors and Windows: Fictional Imagination in Later Chinese Garden Culture.' In *Gardens and Imagination: Cultural History and Agency*, edited by Michel Conan, 191–205. Washington, DC: Dumbarton Oaks, 2008.

Wilson, Ernest H. *China: Mother of Gardens*. Boston, MA: The Stratford Company, 1929.

Xu, Yinong. *The Chinese City in Space and Time: The Development of Urban Form in Suzhou*. Honolulu: Hawai'i University Press, 2000.

Zhou, Xiangpin, and Chen, Zhehua. 'The Shanghai Gardens in Transition from the Concessions to the Present Times.' In *Gardens, City Life and Culture: A World Tour*, edited by Michel Conan and Chen Wangheng, 123–39. Washington, DC: Dumbarton Oaks Research Library and Collection, 2008.

曹林娣。《中國園林文化》。北京：中國建築工業出版社，2005。

———。《蘇州園林匾額楹聯鑒賞》。北京：華夏出版社，2011。

Index

A Taihu stone in Zuibaichi (© 2014 Susie Gordon)

A huang stone formation next to a waterway in Yu Yuan (© 2014 Susie Gordon)

Layering of windows in Zuibaichi (© 2014 Susie Gordon)

Penjing (bonsai) arrangement in Zuibaichi (© 2014 Susie Gordon)

Corridor with inscriptions in Qiuxiapu (© 2014 Susie Gordon)

A moongate in Qiuxiapu (© 2014 Susie Gordon)

Scene from Yu Yuan (© Song Jie, 2015)

The marble boat in Qiuxiapu (© 2014 Susie Gordon)

The dragon wall in Yu Yuan (© 2014 Susie Gordon)

www.ingramcontent.com/pod-product-compliance
Ingram Content Group UK Ltd.
Pitfield, Milton Keynes, MK11 3LW, UK
UKHW020154021126
9729IPUK00034B/106

9 789888 208814